FOSSILS

Description & Interpretation

Within a Biblical Worldview

BY J.D. MITCHELL, MBS

C.E.C.

ISBN 13: 978-0-692-75203-6
ISBN 10: 0-692-75203-X

1 2 3 4 5 6 7 8 9 10

Contents

Photo/Illustration Credits

All photos and illustrations in this book are by the author, unless otherwise noted in the associated caption.

Acknowledgements

I would like to thank my fellow creationist fossil lovers, Otis Kline, Robert Canen, Terry Beh, Dr. Jerry Bergman and Rick Thompson for their numerous suggestions for improving the manuscript. Many thanks go also to my wife, Bonnie, for her assistance with editing and proof reading. However, responsibility for any errors or omissions in this book resides totally with me.

J.D. Mitchell
Gresham, Oregon
December, 2016

CHAPTER ONE

Introduction

Foundational Ideas of the Book

Fossils are remains or traces of life usually found embedded in sedimentary rock. The word "fossil" is from the Latin word *fossilis* which means "dug up."

The foundational idea of this book is as follows: "**Most fossils are evidences of a single year-long cataclysmic event that took place on the earth some 4,500 years ago (the Genesis Flood). These fossils have great value today because they are a reminder of, and proof of, God's judgment, power and grace. This is the Biblical Worldview and it is true. Fossils are not evidences of many events over hundreds of millions of years that have value today because of their great age and their proof of the macroevolution of life. This is the Secular Worldview and it is untrue.**"

A corollary of the book is that Christians should appreciate fossils more than anyone else because fossils:

1. Verify the historical record in the Bible of a worldwide Flood about 4,500 years ago.
2. Are reminders of God's perspective on, and judgment of, evil.
3. Are confirmations of God's actions of grace through

the provision of Noah's ark and the redemptive work of Jesus Christ for sinners on the cross.

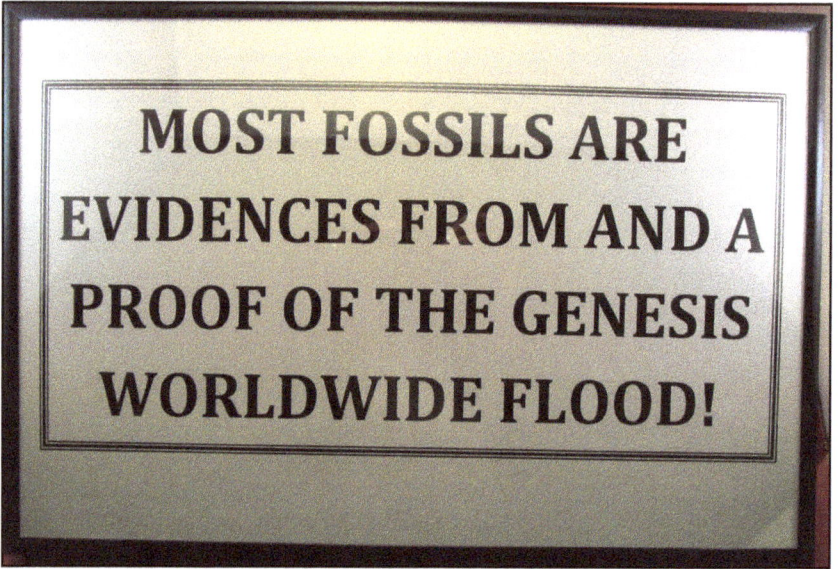

Figure 1: Framed 11" x 17" Sign in Author's Fossil Museum.

Book Theme

The theme that ties the descriptions and interpretations together is the listing of the "Eighteen Facts of Paleontology" that has been compiled and previously published and distributed by the author. This list follows below:

Eighteen Facts of Paleontology:

1. Fossils are almost always formed by rapid and complete burial.

2. Fossils are often found in assemblages (fossil graveyards or bone jumbles).

3. Fossil assemblages indicate catastrophe.

4. Fossils are found in regions all over the earth, even on mountain tops.

5. Most (95%) of the fossils of the world are marine invertebrates (without backbones).

6. The "Geologic Timeline/Column" is a mental abstraction, and is not an observable reality.

7. The sedimentary layers of the Geologic Column have been correlated mostly by fossils.

8. Measurable variation within a kind of lifeform can occur quite rapidly.

9. Variation within a kind has not been, and cannot be, scientifically demonstrated to lead to evolution (change from one kind to another kind).

10. Variation within a kind can be quite extensive, but has definite limits.

11. Biogenesis (life only comes from life) is a law of nature that has never been observed to fail. Abiogenesis (natural life from non-life) has never been observed.

12. All scientists, no matter their worldviews, observe the exact same evidence. Their interpretations of the evidence vary according to their accepted presuppositions.

13. True transitional fossils have not been found, and there is no scientific reason to believe they exist.

14. No lifeform found living or fossilized today is "primitive" or "simple."

15. There is little evidence that fossils are now being formed in lake beds, rivers, or oceans.

16. The present is determined by the past, not vice versa; and

uniformitarianism ("the present is the key to the past") is an assumed principle of naturalism, not a scientific observation.

17. Fossil lifeforms are in many cases identical or very similar to modern lifeforms. These are called "living fossils."

18. Ongoing variation within a kind often results in a reduction of future options for environmental adaptation for that kind.

Each and every fossil listing in this book will reference one or more of these eighteen facts as being relevant to the proper understanding of the description and interpretation of the fossil(s) shown. To get the best use of this book, the reader should place a bookmark at the location of these eighteen facts for ongoing reference.

This fossil guidebook describes and interprets a representative variety of fossils from the rock record using biblical creationist pre-suppositions. Nearly all of the fossils and fossil replicas or casts pictured, described, and interpreted are from the author's personal fossil collection. Therefore, the book also serves as a catalogue for his collection.

This book is unusual and especially valuable to Christians, truth seekers, and skeptics because it presents fossils using biblical creationist presuppositions. Christians and others who accept secular (atheistic) presuppositions about fossils and the rock record will likely misunderstand all of the following:

1. Earth history
2. Human history
3. Geological formations
4. The Genesis Flood
5. The nature of God
6. The nature of man
7. Climate change
8. Biblical creation science
9. The future and eternity
10. God's revelations to man
11. Numerous other aspects of reality

What is in this book?

Any printed-color fossil description guide will be limited to describing only a small fraction of all the fossils that have been discovered in the rock record. There are thousands of different kinds of fossils and hundreds of thousands of specie names have been assigned to life forms by biologists and paleontologists. In *Fossils: Description & Interpretation* the author is displaying and discussing nearly 250 representative invertebrate, vertebrate, plant, and trace fossils from the rock record (at the genus level) that serve to demonstrate the truth of the foundational ideas of this book. At the same time the author is attempting to provide a helpful guide that covers many of the fossil kinds commonly found in secular fossil guides designed for the layman, only with a biblical creationist interpretation.

The fossils described and interpreted are largely limited to macrofossils that can be studied with the naked eye. No microfossils are included (those that require a microscope to see). Also, no nanofossils are included that require an electron microscope to study.

Some fossils found in the earth's crust are not a direct result of Noah's Flood. These fossils resulted from post-Flood activity and are a relative small minority of fossil kinds. None of those fossils from after the Flood are in this descriptive guidebook. Examples include fossils that were the result of the Ice Age that followed the Flood. Representatives of these fossils are a number of mammals, such as mammoths and mastodons. While it is possible that some of these animals, or their parts, were indeed fossilized as a result of the Flood, as well as post-Flood activity, it is the author's opinion that most are not representative of Flood-caused fossils, and so are not described in this book.

What about the Geologic Timeline?

Every one of the many secular fossil guidebooks that the author has inspected places foundational emphasis on the "Geologic

Timeline" or the "Geologic Column." My book theme paleonto-logical fact "number six" states that these are mental abstractions. They exist primarily as the result of deep time presuppositions and the philosophical need to support the anti-biblical idea of mac-roevolution. Even though no true scientific evidence exists that evolution is a fact, without the millions of years inherently sup-posed by the Geologic Timeline, macroevolution is not conceivable for even the most diehard evolutionist.

Therefore, *Fossils: Description & Interpretation* does not include any content based on these atheistic concepts. As the Geologic Timeline is a hypothesis not based on fact, there is no need to accept it; and that is the path consistently chosen for this guide. While some creationists are working on explanations for the Flood time sequence that could replace the secular deep time idea, there is currently no consensus; and the author believes there is currently little hope for a creationist consensus because of the inability for man to fully understand the actual hydraulic, tectonic and vol-canic forces that were in play during the Flood. These forces were orders of magnitude greater than any currently on earth and so are presently outside human understanding. The author believes that Walt Brown's Hydroplate Theory comes the closest to explaining the historical scientific mechanisms of the Flood and that the com-peting theory of Catastrophic Plate Tectonics is not viable.*

The author also believes that creationists who attempt to use the terminology of the evolutionists are working in a counterpro-ductive path. Incorporating the secular era, period, and epoch names of the geologic timeline into any creationist rock record explanation results in an unwanted, and unnecessary, acceptance of many of the atheistic presuppositions contained within the development and usage of those names

---

* See Mitchell, J.D., *Discovering the Animals of Ancient Oregon*, Leafcutter Press, 2013, p. 68-74.

The Biblical Timeline and Fossils

The Biblical Timeline is neither complicated nor difficult to understand. According to the historical narrative of the Bible, God created the universe and everything in it over a period of six ordinary days between 6,000 and 7,000 years ago. Plants and animals were created according to their kinds at approximately the Family taxonomic classification level. Everything was created perfectly until the entry of man's sin not long after the six-day creation was completed. This introduction of sin (the fall) resulted in God instituting the curse that is still present in the remaining imperfect creation in which we live today.

After the fall, man and the creation he was given responsibility for, became more evil until God chose to destroy man with a cataclysmic worldwide Flood some 1,600 years after the creation. Those eight people and air breathing land animals that were on the ark were saved from the Flood. Other kinds of lifeforms that lived in the water or survived on large floating plant matter rafts also made it through the global catastrophe to pass on their genes to descendants, many of which still exist today. Some pre-Flood lifeform variations likely went extinct because of the Flood and others have gone extinct since the Flood for numerous reasons. Variations within kinds to allow adaptation to living conditions, as well as the circumstance of extinction are facts of life that continue yet today. The adaptation capabilities given lifeforms was preconceived as necessary by God, and is possible only because of the information He inserted into the DNA of each living kind in the beginning.

Thus, the vast majority of the lifeforms found as fossils in the rock record are there as a result of the Genesis Flood about 4,500 years ago. Many of the lifeforms of today are identical or nearly identical to those found in the rock record and are called "living fossils." Living fossils cannot be explained logically within the secular worldview, but make perfect sense if the creation and the worldwide Flood happened only thousands of years ago.

Fossil Numbering System

Most of the fossils shown in the images of this book are either real fossils or replica fossils from the author's own collection. The author first had interest as a young boy in dinosaur fossils and his classification system developed from that point. Therefore, his fossil classification numbers have the following prefixes: [DF] for Dinosaur Fossils, [MF] for Mammal Fossils, [MARF] for Marine Fossils, [OF] for Other Fossils, [DR] for Dinosaur Replicas and [OR] for Other Replicas. Each fossil and replica from the author's collection is identified within brackets as such in the image captions. Images of items other than the author's fossils and replicas are identified within brackets with the appropriate information also in the captions.

Types of Fossil Discoveries

Fossils are found in a number of different conditions or states in the earth's crust. The most common conditions are: loose, nodular, inside lamellar sediments, and in consolidated sedimentary rock. After discovery, in any of these conditions, the fossils will need to be stabilized and prepared in some manner for study or display. Some examples of fossils that were discovered in the various conditions are shown in the figures that follow.

Figure 2: Example (l.h,) of a fossil found loose.

Figure 3: Examples of fossils found loose.

Sometimes loose fossils are found lying on the ground and only require a minimal amount of removal of dirt or enclosing matrix. Often they are found in a location that is at an elevation below where they came from and where other fossilized material can be dug out.

Figure 4: Pecten fossil clam in nodule.

Figure 5: Leaf fossil in nodule.

Many fossils are hidden inside nodules of mudstone or sandstone. These nodules are characteristically of a rounded external shape that gives away the possibility of treasure inside. When the nodule is tapped with a hammer, it will often open up into two halves to display the fossil hidden inside.

Figure 6: Fossil fish visible from lamellar sediment splitting.

Sediments laid down during the global Flood will often have fossils trapped between the rock lamina that were formed during the fossil burial and as the sediments solidified. Millions of fossils like these are found, for example, in the Green River formation in Wyoming and the Solnhofen limestone layer of Bavaria.

Figure 7: Dinosaur fossils in very hard rock.

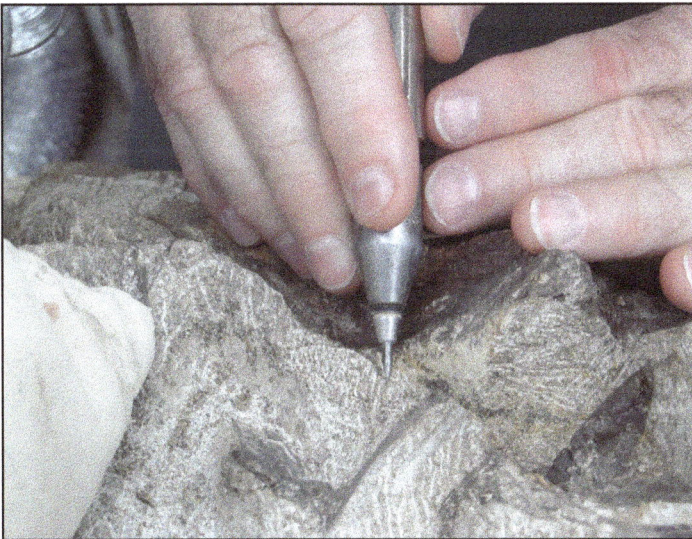

Figure 8: Mechanical removal of matrix around fossil.

When valuable fossils found in hard rock are recovered, it requires the use of specialized tools and lots of tedious hard work. Normally this type of recovery is not attempted unless the fossils are unusually valuable for paleontological study or museum display; or the effort is put forth by workers in a country with very low wages.

Figure 9: Oreodont skull prior to preparation.

Figure 10: Oreodont skull after preparation.

There is great variation in the hardness of the matrix in which fossils are found. Figures 9 and 10 show before and after photos of an Oreodont skull that the author was able to prepare using only dental tools, a brush, and glue because the matrix was not very hard. Vertebrate fossils, especially, are seldom found in an articulated state. The normal condition is to find them totally disarticulated (torn apart). Since only hard parts are easily fossilized by permineralization, the most common vertebrate fossils are teeth, claws, bone fragments and parts of skulls. Museum reconstructions of vertebrate fossil animals are almost always a combination of parts from a number of specimens and locations. The figures below are examples of vertebrate fossils found in "graveyards" and prepared for display in major museums. As can be seen, the mammal fossils in Figure 11 are disarticulated and are held together by the matrix in which they were found. The fossils in Figure 11 represent what would be considered the "best of the best," while those in Figure 12 are more ordinary and better represent the point made by #2 of "The Eighteen Facts of Paleontology."

Figure 11: Oreodont fossils from the Brule formation of the White River Badlands [Houston Museum of Natural Science].

Figure 12: Display of disarticulated vertebrate fossils in graveyard [Denver Museum of Nature & Science].

Authentic versus Replica Fossils

Whereas over ninety per cent of the images of invertebrate, plant and trace fossils in this book are of actual fossils, about fifty per cent of the vertebrate fossils are replica casts of the real fossils. This is because of the scarcity and cost of real vertebrate fossils. As stated by Fact #5 most discovered fossils are marine invertebrates. Vertebrate fossils make up less than one per cent of fossils and plants contribute less than five per cent of fossils.* This scarcity of vertebrate fossils means they are more costly to collect and own. For example, good quality casts of the *Archaeopteryx* skeleton fossils can be purchased for from $150 to $250 each, while the original fossils are valued at many millions of dollars by the museums that hold them. Good specimens of authentic White River formation

---

* Morris, John D. and Sherwin, Frank J., *The Fossil Record*, ICR, 2010, p. 41.

mammal skulls can be worth thousands of dollars, while replica casts can be had for a fraction of the cost. Nevertheless, many fossil collections, including the author's, consist of a high percentage of vertebrate fossils because of the greater interest they generate compared to invertebrate fossils and the fact that vertebrate replicas are usually made of the very best specimens ever found.

CHAPTER TWO

Invertebrate Animal Fossils

Horseshoe Crab *Mesolimulus*

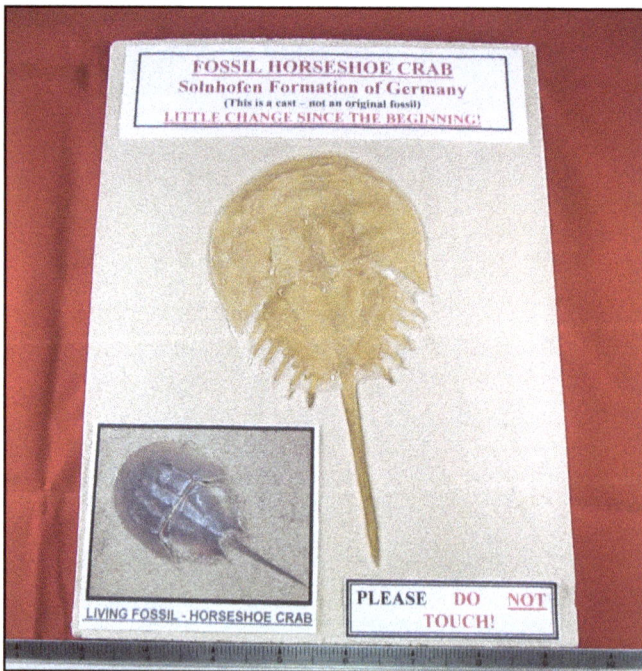

Figure 13: Replica fossil horseshoe crab *Mesolimulus* from the Solnhofen limestone of Bavaria, Germany [OR4].

Description: Relevant "18 Facts" > 1, 6, 14, 15, 16, 17. The replica *Mesolimulus* horseshoe crab is 3 ¾" wide × 7" long in a matrix 7 ¾" wide × 10 ¼" long.

Interpretation: The horseshoe crab is a living fossil, meaning that it is the same in its fossil form as in its living form of today. This makes more sense if the fossils found in the rock record were buried in the cataclysmic worldwide Flood some 4,500 years ago, rather than the fossils having been buried hundreds of millions of years ago according to secular speculations.

There are literally hundreds of living fossil types that have been discovered in the rock record so far and these are strong evidences for the Genesis Flood and against evolution and millions of years. The horseshoe crab is an invertebrate animal which means it has no backbone.

Sea Scorpion *Eurypterus*

Figure 14: Fossil sea scorpion *Eurypterus* from the Bertie Waterline Dolostone group near Utica, New York [MARF66].

Description: The *Eurypterus* tail is 2" long around the curve × ⅜" wide maximum. The matrix is 2 ⅛" wide × 1 ⅞" high × ⅝" thick. Three complete thoracic segments, two partial segments and the tail spike make up this fossil.

Interpretation: Relevant "18 Facts" > 1, 2, 3, 5, 14. *Eurypterus* fossils are found in a number of locations in North America but are common in the Bertie Waterline Group of New York State. As with many fossils they are usually incomplete and the majority represent molts of the animal. These sea scorpions are now extinct but probably lived a life similar to the horseshoe crab.

It makes sense that *Eurypterus* is found with many other marine fossils if it was among the first to be rapidly buried by the actions of the Genesis Flood. Evolutionists consider *Eurypterus* to be primitive, yet it was likely just as complex as a horseshoe crab.

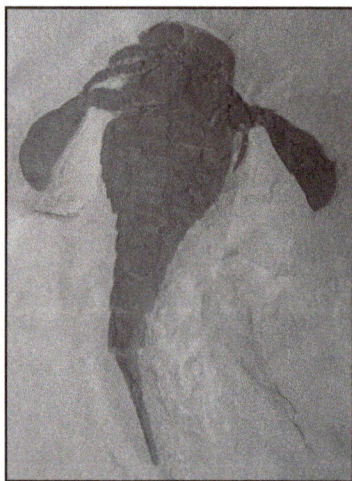

Figure 15: A complete *Eurypterus* fossil from New York that is 5" long [Credit: Viney, M. (2008), The Virtual Petrified Wood Museum, http://petrifiedwoodmuseum.org].

Shrimp *Antrimpos*

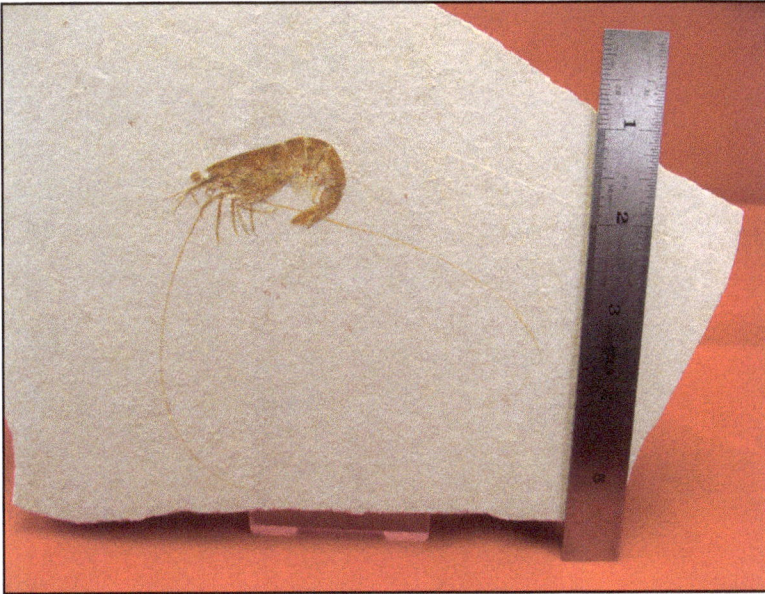

Figure 16: Shrimp fossil *Antrimpos* from the Solnhofen limestone found near Eichstatt in Bavaria, Germany [MARF31].

Description: This aquatic arthropod is from the same area as the famous *Archaeopteryx* fossils. The *Antrimpos* shrimp body is 1 ¾" long × ½" tall. The antennae are about 4" long. The limestone matrix slab is 8 ½" wide × 7 ½" high × ⅝" thick.

Interpretation: Relevant "18 Facts" > 1, 5, 13, 14, 17. The Bavarian limestone quarries are similar to those near Kemmerer, Wyoming, and those in Lebanon, and many very intricate fossils have been recovered from each of these locations. For these fossils to have been preserved, they needed to be rapidly buried in lots of water-borne very fine sediment. Marine invertebrates like this shrimp are the most common fossil type found in the rock record that resulted from the actions of the Genesis Flood about 4,500 years ago.

Lobster *Eryon*

Figure 17: Replica fossil lobster *Eryon* from the Solnhofen limestone of Bavaria, Germany [OR28].

Description: *Eryon* lobster was fossilized upside down and its underside is visible. Two forward claws and eight walking legs with claws are also visible. The fossil is 8 ⅞" long × 5 ⁷⁄₁₆" wide × ⅜" thick. The lobster body is 6 ½" long × 2" wide. Matrix is 12 ⅜" × 8."

Interpretation: Relevant "18 Facts" > 1, 5, 8, 9, 10, 14, 17. According to evolutionists these Solnhofen fossils were made by a process of slow burial over hundreds of years. It is more likely that the reality is that the lobster was rapidly buried. Since this fossil is at least ⅜" thick, it could not have been slowly buried over a period of years without disintegration. The author believes it was rapidly buried in the global Flood at the time of Noah.

The Decapoda order of crustaceans includes crabs, shrimps and lobsters. This complex fossil *Eryon* lobster is undoubtedly an ancestor of the lobsters living today. Differences between *Eryon* and modern-day lobsters are due to variations within the created kind allowed for by their DNA that was designed by God at the creation. God provided for this variation to allow lifeforms to adapt to a variety of changing living conditions. Variation within a kind is not "evolution" of one kind into another kind, and is in reality always limited.

Crinoid *Scyphocrinus*

Figure 18: Fossilized *Scyphocrinus* stalked crinoid (sea lily) from the Sahara Desert, Morocco with two loose fossil crinoid stem sections alongside [MARF32].

Description: The fossil crinoid is 4" wide at the top × 7 ¾" long with a ¼" diameter stem. The matrix is 5" wide × 8" long. There are numerous other crinoid stem sections in the fossil, but of these only one at the top right was made obvious by the fossil preparer and it is ³⁄₁₆" diameter × 1 ¹⁄₁₆" long.

Interpretation: Relevant "18 Facts" > 1, 2, 3, 5, 14, 17. Crinoids are flower-like (thus the name "sea lilies") often beautifully colored invertebrate echinoderms, which grow in colonies on the sea floor. There are many genera identified in the rock record and in the seas of today. All have a five-fold radial symmetry, but vary in shape, plates and arms.

"When crinoids die they disarticulate fairly rapidly, for within a few days of death the ligaments and soft tissues which hold the component skeletal elements together have decayed. Where a crinoid has been preserved intact (or nearly so) it would have been buried rapidly and with sufficient cover to avoid disarticulation of the component parts…"*

Figure 19: Fossilized root system for a crinoid
[Wyoming Dinosaur Center].

---

*   Taylor, Paul D. and Lewis, David N., *Fossil Invertebrates*, Harvard University Press, 2005, p. 173.

The above quote by evolutionary scientists matches the biblical creationist view, not only about crinoids, but also about the vast majority of other fossils. Crinoids were created in the beginning, many were buried in the Genesis Flood to be found later as fossils, and many still live today in the seas of the earth.

Blastoid *Pentremites*

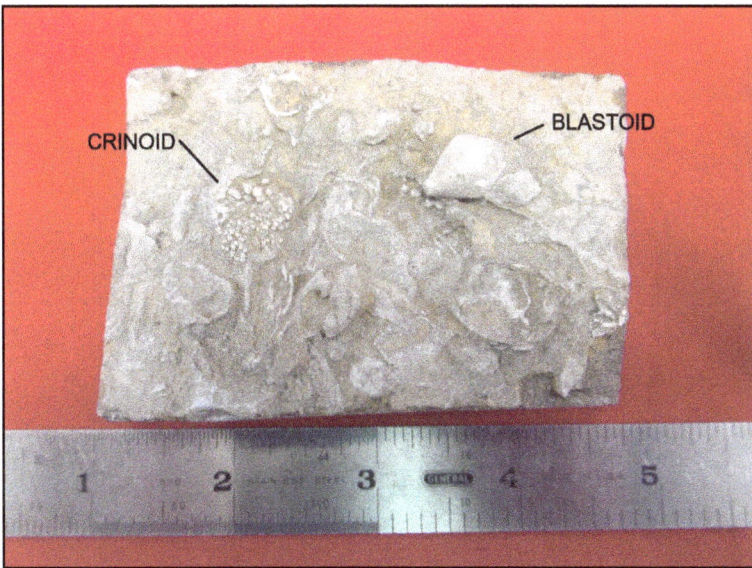

Figure 20: Fossil blastoid *Pentremites* and parts of a crinoid (perhaps *Taxocrinus*) from the Bangor Limestone formation of northern Alabama [MARF54].

Description: The blastoid rosebud is ⅜" diameter × ⁹⁄₁₆" long and the crinoid remnant is ½" diameter. The limestone matrix is 2 ⅛" wide × 3 ¼" long × 1" thick. The two identifiable fossils are embedded in a mass of other invertebrate fossil pieces.

Interpretation: Relevant "18 Facts" > 1, 2, 3, 4, 5, 14. Blastoids are small extinct echinoderms that were attached to the sea floor by a thin stem in a manner similar to the crinoids. The top "flower-like portion" is called the theca and often reminds one of a rosebud.

The biblical creationist asks, "Why do fossils like these not regularly form today?" If they did the ocean floor would be a conglomerate of fossils miles deep. It makes more sense that these fossils formed due to an unusual cataclysmic event that the Bible describes as a worldwide Flood.

Bryozoan *Fenestella*

Figure 21: Fossil bryozoan *Fenestella* in matrix from the Bangor Limestone formation in northern Alabama [MARF55].

Description: There are a number of disconnected lacy netlike pieces of *Fenestella* in the limestone matrix. The zooid nets consist of tiny squares about $\frac{1}{64}$" in size. The stem-like pieces may be the spines that rooted the colonies to rocks or shells on the ocean floor. The matrix is 2 ½" wide × 3 ½" long × ¾" thick.

Interpretation: Relevant "18 Facts" > 1, 2, 3, 5, 11, 14. Bryozoans are miniature colonial animals consisting of numerous connected individuals called zooids. Today they are found living in sub-tropical to cold water environments, but many of the bryozoan types found fossilized look to be extinct. Such is the case for the *Fenestella* genus. Bryozoans look like some corals but are more complex having nervous, muscular and digestive systems.

These tiny animals can be seen with the naked eye but are best studied under a microscope in order to see the fine detail of their construction. When man-made things are examined with microscopes, they lose detail in proportion to the amount of magnification. God's creations, on the other hand, continue to show wonderful design with the largest magnifications. The Law of Biogenesis and common sense tell us that even these relatively simple animals have great complexity and did not evolve from inorganic materials.

Bryozoan *Archimedes*

Figure 22:
Fossil bryozoan *Archimedes* in matrix from the Bangor Limestone formation in northern Alabama [MARF56].

Description: The *Archimedes* structure is 3 ⅜" long × ¼" maximum diameter. There are 15 visible threads in the fossil. The matrix is 2 ½" wide × 5 ¼" long × ¾" thick.

Interpretation: Relevant "18 Facts" > 1, 2, 3, 5, 11, 14. This wonderful bryozoan animal was undoubtedly named after the Archimedes screw developed by man thousands of years ago to pump irrigation water. Many human inventions man have been inspired by the works of God found in nature. It is possible that an *Archimedes* bryozoan or another of God's similar designs inspired the first Archimedes screw.

Bryozoan *Constellaria*

Figure 23: Six fossil specimens of the bryozoan moss animal *Constellaria* from the State of Kentucky [MARF68].

Description: All six specimens have hundreds of bumps (monticules) that are characteristic of the *Constellaria* bryozoan genus from Kentucky. Specimens tend toward the branching design rather than the massive or encrusting types. The feeding zooids would have lived in the monticules. The largest bryozoan in the photo is 2" × 1 ³⁄₁₆" × ³⁄₁₆" thick. The smallest is 1" × ⁷⁄₈" × ³⁄₁₆" thick.

Interpretation: Relevant "18 Facts" > 1, 2, 3, 5, 8, 9, 10. *Constellaria* genera from other areas of the world are somewhat different from the ones shown here from Kentucky. They often have star-shaped monticules. This is an indication of variation that occurred within this bryozoan kind between the time of the creation and the Flood.

Sponge *Astylospongia*

Figure 24: Three fossil sponges *Astylospongia* from the Beech River formation in Tennessee [MARF70].

Description: All three sponge fossils are basically spherical in shape. They are all about 1" in diameter. Some of the internal structure can be seen in the center specimen.

Interpretation: Relevant "18 Facts" > 11, 14, 17. Evolutionists characterize sponges like *Astylospongia* as simple or primitive animals. However, man cannot make sponges that would live as do the ones created by God. Sponges are very complex with multicells of different designs that can capture food particles, derive nourishment from the food, and then expel the waste. They must have some sort of skeletal system as well, usually consisting of complex needlelike spicules.

The reason for the emphasis by evolutionists on the idea

of "simple" life is that they must convince the gullible that life somehow came from non-life. However, the Law of Biogenesis specifies that life can only come from life. This is called a law because it has never been observed to fail. We know that some types of sponges still are alive today, and probably have always existed as such since creation.

Sponge *Hindia*

Figure 25: Fossil sponge *Hindia* from the Beech River formation in Henderson County, Tennessee [MARF71].

Description: The *Hindia* fossil sponge is nearly spherical and is about 1 ¼" × 1 ⅝."

Interpretation: Relevant "18 Facts" > 11, 12, 13, 16, 17. Evolutionists have placed sponges in the phylum Porifera where they have not been able to identify any common ancestors. That is, they believe sponges have existed ever since the "explosion of life" (Cambrian) with their supposed ancestors not found anywhere in the rock record.

The *Hindia* fossil sponges have left no evidence that they were attached to anything while alive like so many other sea animals. Perhaps they were designed to roll freely about on the sea floor. At any rate, the fossil record for sponges aligns with the biblical creationist view that sponges were created by God in the beginning to be sponges.

Graptolite *Phyllograptus*

Figure 26: Shale matrix with seven or more stipes of the graptolite fossil *Phyllograptus* from the Toyen formation near Oslo, Norway [MARF83].

Description: The longest *Phyllograptus* stipe is 1 ¾" long and the widest stipe is ¼" wide. The shale matrix is 5 ½" wide × 3 ¹¹⁄₁₆" high × ½" thick. The tubular stipes are all flattened onto the shale, but thecae can clearly be seen with a magnifying glass.

Interpretation: Relevant "18 Facts" > 1, 2, 3, 5, 13, 14. Graptolites are fossilized marine colonial organisms with each colony having one or more tubular branches (stipes). It is thought that most of these extinct organisms floated on the oceans among the plankton they are thought to have fed on. The stipes bore numerous cup-like structures called thecae that housed the individual animals called zooids, similar to those in corals and bryozoans.

Evolutionists consider graptolites to be part of their imagined long macroevolutionary trek of "goo to you." That is, they consider graptolites as in the evolutionary line to the vertebrates. But there is no known common ancestor for them and since they seem to be extinct, no evolutionary link to the next advanced step of life.

The biblical creationist could hypothesize that graptolites were created by God in the beginning as graptolites, and then went extinct due to conditions during and/or after the global Flood. There is no way to experimentally test any extinction hypothesis for these graptolites because the conditions of the Flood cannot be repeated today.

Graptolite *Monograptus*

Figure 27: Shale matrix with over a dozen stipes of the graptolite fossil *Monograptus* from the Bardo syncline, Holy Cross Mountains, Poland [MARF87].

Description: The longest *Monograptus* stipe is 2 ¼" long and the widest stipe is ¹⁄₁₆" wide. The matrix is 7 ½" long × 4" wide × ¼" thick. All stipes are straight with theca visible to one side only (with magnification).

Interpretation: Relevant "18 Facts" > 1, 2, 3, 5, 13, 14. The *Monograptus* genus of graptolite fossil is always found with a single branch and with thecae running on one side only. However, *Monograptus* stipes can be found straight, gently curved, or even spirally twisted. These fossils are geologically important because they are defined as index fossils.

To the biblical creationist, index fossils can be useful for helping to determine which types of fossils, minerals, or elements are often found together in the sedimentary layers of the rock record. While the secular scientist may think an index fossil determines the age of a particular batch of rocks, this is not reality. Instead it is only circular reasoning because the ages of the index fossils are *a priori* assumed to be of a certain age according to the geologic timeline which is simply a mental abstraction. The most common index fossils are marine invertebrates for all stratigraphic layers, which makes sense because marine invertebrates make up the vast majority of all fossils.

Graptolite *Didymograptus*

Description: The longest *Didymograptus* stipe is 1 ¾" long and the maximum stipe width is ⁵⁄₃₂" wide. All thecae run along the insides of the stipes and are shaped like saw teeth. The angle between the stipes is 15 degrees on the left specimen and 30 degrees on the right specimen. The matrix is 3 ½" wide × 3 ⅝" high × ¼" thick.

Interpretation: Relevant "18 Facts" > 1, 2, 3, 5, 13, 14. *Didymograptus* is identifiable by its tuning-fork shape, but it is not known how the extinct graptolite colonies would have oriented themselves in the ocean waters. It is also not known how the zooids that are assumed to have inhabited the thecae looked or operated. The location on

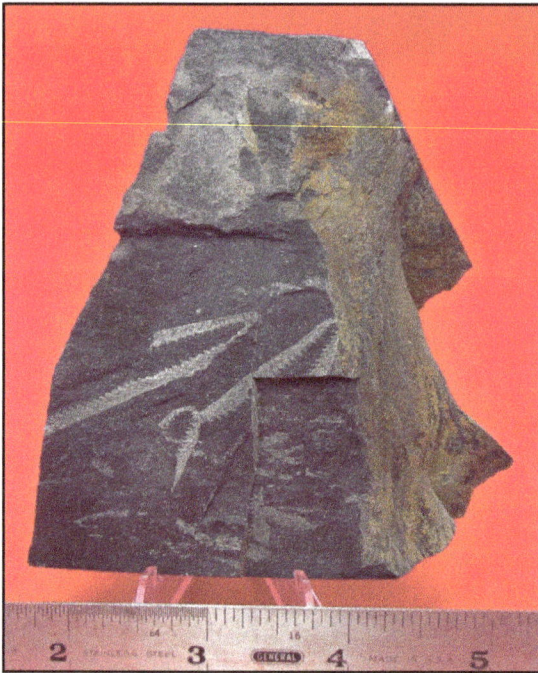

Figure 28: Shale matrix with two nearly complete specimens and some other parts of the graptolite *Didymograptus* from Dyfed, Wales, United Kingdom [MARF88].

each graptolite where the stipes originated is called the sicula. Experts believe that while the stipes and theca are always flat as fossils, they must have been 3-dimensional when living.

Since graptolites are currently understood to be extinct, it is not possible to know as much about them as would be possible if they were still living. Nevertheless, to the biblical creationist their complex and amazing design indicates a Designer rather than a common ancestor.

Bivalve *Katherinella*

Description: The conglomerate is held together by siltstone and is 5" long × 2 ½" wide × 3 ½" high. The larger clam shells in the matrix average 1 ¾" long × 1 ¼" high. The seven loose fossil clams are all complete with both valves (clamshells) frozen tightly

Figure 29: *Katherinella* fossil bivalve clams. The conglomerate is from the State of Washington. The seven loose fossil bivalves are from the Astoria formation near Newport, Oregon [MARF27].

together and average about 2" long × 1 ¾" high. Some of the loose clams show the effects of wear from the surf prior to collection.

Interpretation: Relevant "18 Facts" > 1, 2, 3, 5, 7, 14, 17. Along the Pacific Coast of Oregon at a number of locations can be found many fossil bivalve clams like these that have been eroded out of the banks. A high percentage of the *Katherinella* are found with their two valves shut tightly together. On the other hand, similar extant dead clams always pop apart into an open clam or even two separate valves lying on the beach. This indicates that the fossil clams must have been rapidly buried while they were still alive!

The author has noticed that many of the fossil clams from the rock record have different genus names from extant clams that look very similar. In some cases the fossil clams are so identical to living clams that they have been assigned the same genus names and are called "living fossils" by the experts because they are thought to have lived millions of years apart.

Bivalve *Pholadomya*

Figure 30: *Pholadomya* fossil bivalve clam from Blockley, Gloucestershire, United Kingdom [MARF17].

Description: The *Pholadomya* fossil clam is 2 ½" long × 1 ⅝" high × 1 ¾" thick. Both valves are complete and totally permineralized with the valves clamped tightly shut.

Interpretation: Relevant "18 Facts" > 1, 5, 7, 17. This particular genus has been found in many locations throughout the world and is alive yet today, so is another example of a living fossil. The *Pholadomya* fossils are commonly discovered with the valves tightly closed like this specimen, indicating rapid burial while the clam was still alive. In the rock record it can be found throughout many layers comprising the secular "Geologic Column" yet it is an index fossil for specific sedimentary layers thought by evolutionists to be millions of years old. One geographical area familiar to the author where this fossil clam is found, and is used by evolutionists as an index fossil, is in the central part of Oregon State.*

---

* See Mitchell, J.D., *Discovering the Animals of Ancient Oregon*, Leafcutter Press, 2013, p. 241.

Bivalve *Anadara*

Figure 31: Bivalve clam fossil *Anadara* from the Astoria formation near Newport, Oregon [MARF79]

Description: The fossil clam is 1 ⅞" wide × 1 ⅝" high. The left portion is broken away and is missing, however there is little wear or abrasion on the exterior of this fossil. Five growth rings are visible.

Interpretation: Relevant "18 Facts" > 1, 2, 3, 5, 17. According to some sources *Anadara* is Oregon's most common marine fossil.* It looks very similar to cockles that inhabit our oceans today. Again, we see a bivalve that was fossilized with both valves tightly closed, indicating rapid burial while the clam was still alive – an evidence for the worldwide Flood.

---

* For example see, NOAA – Sea Grant Oregon, "Fossils You Can Find on Oregon Beaches," 2004.

Bivalve *Cerastostreon*

Figure 32: Fossil oyster shell *Cerastostreon* from the Walnut formation in Bell County, Texas [MARF72].

Description: The oyster fossil is 2" wide × 2 ½" long × ¾" thick. There are about 12 irregular ridges on the exterior of the valve (shell) in a spread fan pattern.

Interpretation: Relevant "18 Facts" > 12, 14, 17, 18. *Cerastostreon* oysters are bivalve clams and have two valves that are mirror images of each other. The exteriors of the valves are so distinctive that even a novice seafood buyer can identify them.

The rock record shows that oysters have always been identifiable as oysters, but with much variation within the oyster kind. Oyster fossils are found in the rock record all over the world. Extant oysters are also very common, in part because of intentional plantings for food as humans have spread throughout the earth.

Oyster *Lopha*

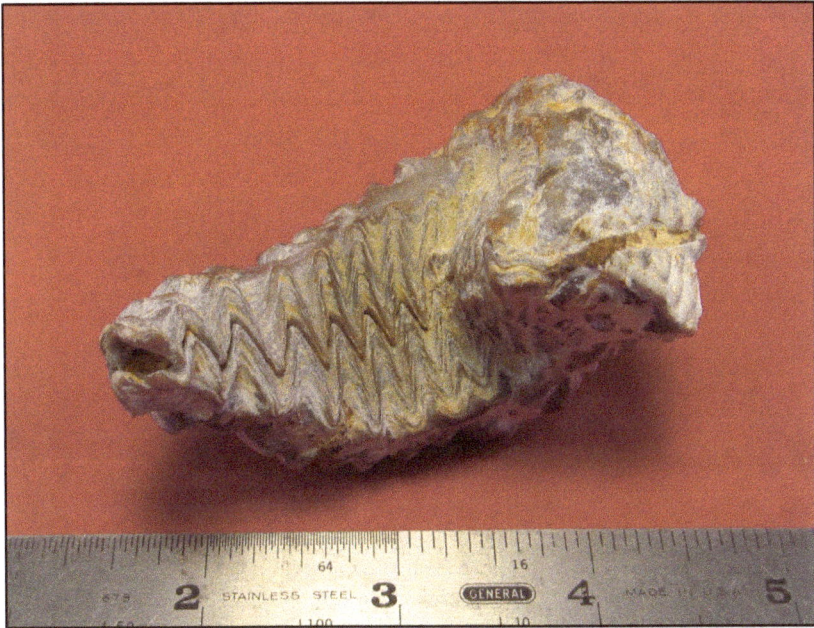

Figure 33: Bivalve oyster *Lopha* from the Morondava River Basin of Madagascar [MARF86].

Description: The fossilized *Lopha* oyster is 3" end-to-end × 1 ½" wide × 1 ½" maximum thickness. The zig-zag saw teeth vary from ⅛" to ¼" on centers and the tooth heights vary from less than ⅛" high to over ¼" high. The shell is prominently ribbed with 20 ribs on the convex (back) side and six ribs on the concave (visible) side. The valves are not equal in width.

Interpretation: Relevant "18 Facts" > 8, 9, 10, 17, 18. The *Lopha* animals living and fossil show amazing design with considerable shell variation. For this reason, they have common names, such as zig-zag oyster, cockscomb oyster, and saw tooth oyster. The fossil specimen in the figure above from Madagascar is locally called a zig-zag oyster. *Lopha* oysters are still found alive today, thus it is a living fossil. Living *Lopha* are stationary suspension type feeders and filter sea water to extract nutrients. They are found mostly in tropical and subtropical ocean waters.

These *Lopha* oyster fossils from Madagascar are examples of where human tools (such as saw teeth and screw threads) were first invented by the Creator. The biblical creationist sees many hallmarks of design in living things that the evolutionist often misses because of his incorrect presuppositions that everything came from nothing for no reason, and animals and plants are able to invent things themselves. Creationist books document many of these designs which they attribute to their true Designer.*

Bivalve *Pecten*

Figure 34: Partial fossil scallop *Pecten* from the Empire formation of Coos County, Oregon [MARF57]

Description: The scallop is 3 ⅞" wide × 3 ¼" high × ¾" thick. There are 20 remaining ribs, but both ears of the valves and the hinges are missing. The valves are also damaged on the ends that open.

---

* For example see: Burgess, Stuart, *Hallmarks of Design*, One Day Publications, 2004.

Interpretation: Relevant "18 Facts" > 1, 4, 5, 14, 17. Pecten scallops are found alive worldwide today as well as fossils in the rock record. The fossil scallop pictured above was buried rapidly when alive as shown by the fact that both valves are permanently closed tightly. That is not how scallop shells are found today on the ocean beaches. If an extant animal has died the hinge ligaments that open and close the valves always relax so that the valves will then pop open. Rapid burial by massive amounts of water-laden sediment best explains this fossil which is good evidence for the Genesis Flood.

Bivalve *Chesapecten*

Figure 35: Three fossil scallop valves *Pecten* or *Chesapecten* from the Yorktown formation (James River), Virginia [MARF64-1, 2, 3], and a scallop seashell collected from a beach at an unknown location and sold in an Oregon gift shop (far right).

Description: All four of these specimens range from 2 ⅝" to 3" wide × 2 ¾" to 3" high. Fossil-1 has 11 exterior reinforcing ribs, fossil-2 has 10 ribs, fossil-3 has 15 ribs, and the recent seashell has 23 ribs. The hinges show nearly identical design for all four specimens and the amount of shell curvature is nearly the same for all.

Interpretation: Relevant "18 Facts" > 5, 8, 9, 10, 13, 17, 18. This photo illustrates variation within a kind. When God created the scallop kind He designed its DNA to allow for the variation seen in the photo. The three fossils are dated by evolutionists to be many millions of years old, but are really closer to 4,500 years old according to biblical creationist presuppositions.

Notice that the ribs provide strength to the shells. Every mechanical and civil engineer knows that a ribbed homogeneous plate is many times stronger in bending than one without ribs. Therefore, the DNA variation provided for scallops is not likely to allow for *Pecten* scallops to change to having no ribs, because to do so would result in a shell that could not withstand the external forces the shell normally experiences in its habitat.

Bivalve *Gryphaea*

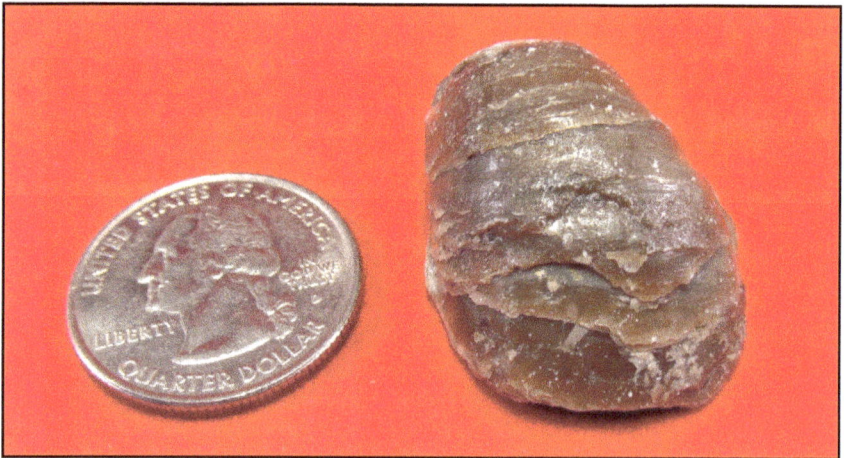

Figure 36: Fossil pelecypod *Gryphaea* popularly called the Devil's Toenail [MARF62].

Description: The *Gryphaea* fossil is 1 ¼" long × 1 ¹⁄₁₆" wide × ¹¹⁄₁₆" thick. There are eight growth ridges visible on the exterior surface. This fossil represents half of the animal's protective shell and is the larger left valve. The right valve fits inside the left like a lid.

Interpretation: Relevant "18 Facts" > 1, 4, 8, 10. It is easy to understand why this pelecypod bivalve animal's shell is called the Devil's Toenail. The specimen in the photo is somewhat smaller than those typically found. Most are about 2 ¼" long with the largest up to 3 ½". The discovery location of the fossil pictured has been lost, but Devil's Toenails are found worldwide.

A number of fossil pelecypods similar in design to *Gryphaea* are known but have been assigned different genus names. This variation could have occurred during the 1,600 years from the fall to the Flood. For most creation scientists, the created kind is established at, or near, the Family level rather than the Genus level.

Brachiopod *Spirifer*

Figure 37: *Spirifer* fossil brachiopod from Ohio, USA [MARF34].

Description: This beautifully formed and permineralized fossil is 2" wide × 1 ³⁄₈" deep. About 16 rounded ribs are on each side of the top valve and a similar number on the bottom valve. Flakes of various sizes of pyrite are scattered over the entire surface of the brachiopod.

Interpretation: Relevant "18 Facts" > 1, 5, 14. In Genesis chapter six we read the reasons why God destroyed the entire surface of the earth with the global Flood some 4,500 years ago. Man's wickedness and the evil of his heart resulted in the destruction of all humans and land animals except for those on Noah's ark. In addition, the surface of the earth was greatly changed. However, in God's grace the post-Flood earth still had many beautiful things that we can enjoy today including this wonderful *Spirifer* brachiopod that sparkles in the light.

Pyrite (iron sulfide) is often called "fool's gold" because it looks so much like real gold. But people who appreciate the beauty and design of a multitude of things in the universe are not fools if they attribute them to the true Creator. Those who reject the Creator are indeed foolish and without excuse (Romans chapter one).

Brachiopod *Platystrophia*

Figure 38: Twenty representative *Platystrophia* brachiopod fossils from several locations in the State of Kentucky [MARF 67].

Description: The five *Platystrophia* in the top row show the convex valve sides while the 15 in the lower three rows show the concave valves sides. All the brachiopods are strongly ribbed with sharp-crested ribs at 16 to 20 ribs per side. They range in size from ½" wide to 2" wide. The complete brachiopods are about two-thirds as thick as they are wide. All but three of the brachiopods are fossilized with both valves tightly shut.

Interpretation: Relevant "18 Facts" > 1, 2, 3, 5, 8. These brachiopods are usually found in the lowest fossil-bearing sediments, indicating that these animals lived at the bottom of the oceans. In a manner similar to many other sea animals with opposing valves, brachiopods are often found fossilized with both valves tightly closed. This is an indication of rapid burial while the animals were still alive.

The condition of these twenty specimens is from nearly perfect shape and symmetry to quite distorted and/or broken. This can best be explained as largely due to variation within a kind and the effects of the forces on the animals from the worldwide Flood that buried them.

Gastropod *Archimediella*

Figure 39: Fossil gastropod *Archimediella* sometimes identified as *Turritella* (left) and an Indian Auger shell (right) [MARF43].

Description: Archimediella is 1 ¹⁄₁₆" diameter on the large end × ⅛" diameter on the small × 3 ⅛" long. There are ten whorls each having numerous cord lines with the number of lines increasing with whorl size. The aperture is broken off slightly. The Indian Auger shell is ¹¹⁄₁₆" diameter on the large end × ⅛" diameter on the small end × 3 ⅛" long along the axis. There are 12 whorls each with cord lines similar to the fossil gastropod. The aperture of the shell is thinner than the one in the fossil and is not broken.

Interpretation: Relevant "18 Facts" > 1, 5, 8, 9, 10, 17. There are scores of these coiled gastropod varieties found not only as fossils but also found alive today in the oceans and on the beaches. If the fossil is permineralized as in the case of the *Archimediella* seen here it will be considerably heavier than the shell of a recent similar gastropod found on a beach.

The living examples of these marine animals all have a head with eyes and a mouth along with a flattened foot for crawling. It can be assumed that the fossilized shells represent animals that were constructed similarly. These gastropods are wonderful

Figure 40: Mass assemblage of *Turritella* from France [Wyoming Dinosaur Center].

examples of the tremendous variety within a created kind as allowed for by the information in their DNA.

Gastropod *Goniobasis*

Figure 41: Blocks of *Goniobasis* fossilized gastropod snails in a matrix of chalcedony sedimentary rock [OF2].

Description: Most of the encased hundreds of *Goniobasis* gastropods are from ¼" long to 1" long. The matrix exhibits three different colored layers: white, gray and brown. The large block is 3" × 3 ½" × 5" and weighs three pounds. The small piece is 1" × 1 ¼" × 3".

Interpretation: Relevant "18 Facts" > 1, 2, 3, 4, 5, 8, 9, 10, 14, 17. This is an example of how confusing the naming of some fossils can be. For years most rock shops have identified this conglomerate of snail fossils encased in a cryptocrystalline variety of quartz as "Turritella Agate." The reason is that the snails look very much like the marine fossil *Turritella*. But they are not the same as *Turritella*

Figure 42: South Dakota rock shop container of hundreds of pounds of *Goniobasis* chunks identified as Turritella Agate.

and are actually properly named *Goniobasis* or (according to some experts) *Elimia*.

Whatever the name, the fossil-stone is mined by the ton from quarries in Wyoming, Colorado and Utah and then sold by the pound. Many of these pieces are then sliced, tumbled and polished into various items of beauty such as rings and bracelets.

As with the *Turritella* fossils there is considerable variety within these *Goniobasis* snails. If these "Turritella Agate" fossilized sedimentary blocks were the result of catastrophic rapid burial, then it would be difficult to determine if they were freshwater or marine or if they could have adapted in their various forms to both types of water.

Chalcedony is a variety of quartz, and quartz is crystalized silica which is an important material for forming both rocks and fossils. Since *Goniobasis* is found in numerous sedimentary layers of the rock record, that indicates that there was a lot of mixing of the habitat of these snails during the global Flood.

Gastropod *Liracassis*

Figure 43: Gastropod (snail) fossil *Liracassis* from the Astoria formation near Newport, Oregon [MARF80].

Description: The snail is 1 ¾" diameter × 2 ⅝" long. Portions of the whorls are worn away and the exterior is covered in a layer of hard matrix in other places. The shell looks to have had 20 to 25 whorls prior to being buried when *Liracassis* was alive.

Interpretation: Relevant "18 Facts" > 1, 5, 15, 17. The *Liracassis* fossil gastropod looks much like the *Tonna* giant tun shells found today in tropical oceans. The specimen above has evidence of having been moved from its original burial location after fossilization.

Ammonite *Desmoceras*

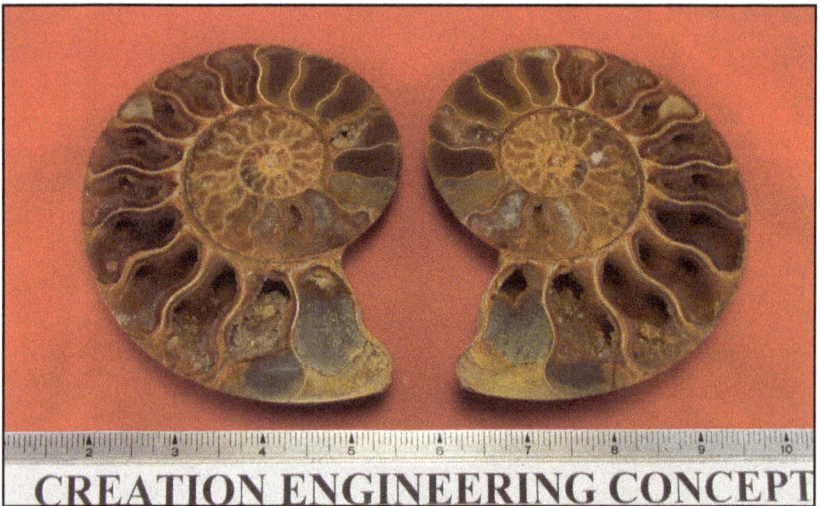

Figure 44: Cut and polished fossil ammonite from the island of Madagascar [MARF26].

Figure 45: Exterior view of fossil ammonite in Figure 44 [MARF26].

Figure 46: Close-up photo of cut and polished fossil ammonite showing calcite crystals and interior detail including siphuncle [MARF 26].

Description: Each half of the spiraling ammonite is 5 ³⁄₈" wide × 4 ¼" high × ¹¹⁄₁₆" thick maximum. The largest chamber at the spiral end is ¾" wide × 1 ⅝" long. There are 50 countable septum-separated chambers in the fossil. All septae are curved to various shapes with none being straight. Calcite crystals are visible in all of the chambers. The fossil is completely permineralized and sutures are visible on the left half of the exterior view. Matrix material has not been removed from either half at the exterior center of the ammonite. The siphuncle is located at the edge of the spiral unlike in the nautilus where it goes up the center. (The siphuncle is a thin-walled tubular extension of the mantle that passes from chamber to chamber and regulates the buoyancy of ammonites and nautiloids.)

Interpretation: Relevant "18 Facts" > 1, 2, 3, 5, 6, 7. Fossil importers throughout the world love to sell these beautiful fossils from Madagascar that have been sliced in half and then polished. Many of the smaller specimens are made into earrings or pendants. Usually these ammonite fossils are identified with genus names *Cleoniceras* or *Desmoceras*.

Nearly all of the ammonite fossils from Madagascar have been somehow altered for added value because Madagascar export laws make it illegal to export unprocessed fossils. Ammonites have been found in locations throughout the world and often take on many beautiful minerals through the permineralization process. Ammonites are one of the fossils often used as "index" fossils by secularists to conform to their geologic timeline. In reality all of these fossils were rapidly buried sometime during the year-long worldwide Flood.

Ammonite *Cleoniceras*

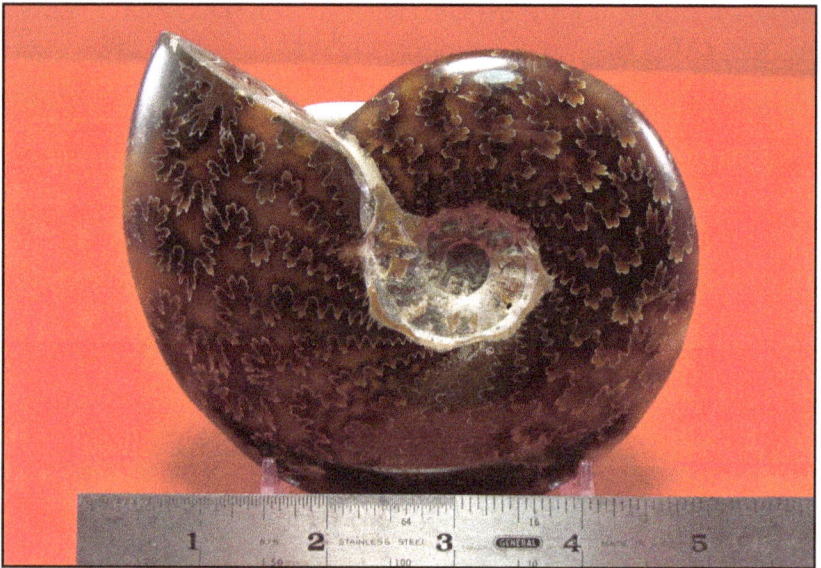

Figure 47: Whole sutured *Cleoniceras* fossil ammonite from Madagascar [MARF37].

Description: The *Cleoniceras* fossil ammonite is 4 ½" wide × 3 ⅞" high × ¹⁵⁄₁₆" thick and has had a large portion of the in-filled surface polished by island craftsmen. Complex suture lines are prominent over the entire surface.

Interpretation: Relevant "18 Facts" > 1, 2, 3, 5, 6, 7. "A suture line is the pattern made by the lobes and saddles of a single septum around its entire edge. Generally the suture line is simple for the first few septae, and becomes more complex with successive ones. Groups of closely related ammonite species often have similar suture lines, and so the suture line is useful in classifying ammonites."*

The varying beautiful suture patterns on ammonites are best explained as ramifications of God's artistic and engineering capabilities. How they could vary from genus to genus due only to evolutionary processes requires imagination beyond that of the author. As with other fossils from Madagascar, human effort is needed to bring out the beauty inherent in these extinct animal fossils from the Flood.

Ammonite *Craspedites*

Figure 48: *Craspedites* fossil ammonites from Russia. Upper specimen is from the Volga River and the Ulyanovsk Region [MARF15]. The lower cut and polished specimen is from Saratov, Russia [MARF16].

---

* Monks, Neale & Palmer, Philip, *Ammonites*, Smithsonian Institution Press, 2002, p. 153.

Description: Upper specimen with sutures is ⅞" across × ⅜" thick maximum and is connected to a chunk of matrix ⅞" × ⅞" × 1 ¼". The chambers on this *Craspedites* specimen are preserved as grayish-white calcite crystals and the septa are preserved as black calcite. The cut and polished specimen is 1 ½" in size and has a typical ammonite chamber and septa design.

Interpretation: Relevant "18 Facts" > 1, 4, 5, 14. The value of these specimens is to illustrate ammonite beauty and design; and the upper specimen helps in understanding the connection between the septa that separate the chambers and the suture marks left by their connection to the inside of the shell. This beauty and design is easier to understand as coming from the mind of God than from the random interaction of matter over deep time.

Ammonite *Dactylioceras*

Figure 49: Slab containing many fossil ammonites *Dactylioceras* from Holzmaden, Germany [MARF11].

Figure 50: Mass assemblage of *Dactylioceras* ammonites from Germany [Wyoming Dinosaur Center].

Description: There are at least twenty *Dactylioceras* ammonites of various sizes embedded in the matrix of [MARF11], many requiring magnification to properly study. The largest specimen is 1 ½" across and the triangle-shaped specimen is 5 ¾" high × 7 ¼" wide × ½" thick.

Interpretation: Relevant "18 Facts" > 1, 2, 3, 4, 5. These images are of what secular scientists describe as "mass mortality plates." There are innumerable fossils jammed together and piled one on the other in these fossil jumbles. The interpretive plate at the Wyoming Dinosaur Center display explained the jumble as the result of a storm, but storms regularly happen today without the same results. These types of fossil graveyards match up nicely with the expectations of the conditions caused by the Genesis Flood.

Figure 51: Ammonite fossil *Dactylioceras* from New Yorkshire, England [MARF36].

Description: The *Dactylioceras* is 1 ¼" wide × 1 ⅜" high × ⅜" thick. The ammonite is embedded into a matrix on the back for a total thickness of ⅝". This fossil is pyritized.

Interpretation: Relevant "18 Facts" > 1, 2, 3, 4, 5. Pyritized *Dactylioceras* is best known from northern England. Many of the fossils from there have had snake heads carved into the openings so they can be sold to unsuspecting tourists as petrified coiled fossil snakes. It has an evolute shell in which successive whorls slightly overlap. The variations in the color and hardness of the *Dactylioceras* fossils in the three figures above are due to the differences in the minerals that were adjacent to the ammonites when or after they were buried in the Flood.

Ammonite Reconstruction

Figure 52: Reconstruction of what a typical ammonite animal may have looked like when alive [Houston Museum of Natural Science].

Ammonite *Promicroceras*

Figure 53: Fossil ammonite *Promicroceras* from Dorset, United Kingdom, Lyme Regis [MARF12].

Description: Six ammonites are visible in the sandstone matrix along with several casts of others. The largest ammonite is ⁹⁄₁₆" wide. The matrix is 2 ½" high × 3 ½" wide × ⅝" thick.

Interpretation: Relevant "18 Facts" > 1, 4, 5, 14. There are two different styles of ammonite in this small rock slab. The ribbed ammonites are genus *Promicroceras*, an ammonite commonly found along the Dorset Coast of England.

Ammonite *Perisphinctes*

Figure 54: Fossil ammonite *Perisphinctes* from France [MARF7].

Description: This fossil specimen of *Perisphinctes* is 2 ¾" wide × 2 ⅜" high × ¹¹/₁₆" thick. The shell is involute with over 100 fine branching ribs and the chamber sections are visible nearly from the beginning.

Interpretation: Relevant "18 Facts" > 1, 2, 3, 4, 5, 14, 15. According to all accounts, ammonites are extinct and millions of them are found as fossils in sediments all around the earth. Yet there is no indication that fossils are being formed to any noticeable extent under any conditions anywhere today. This is strong evidence that most fossils are a result of the worldwide Flood at the time of Noah.

The *Perisphinctes* ammonite is used by evolutionists as an "index fossil" to identify the age of sediments in which they are found. However, the author has noticed that any index fossil which does not match the secular preconceived notion of rock age is ignored or renamed.*

Many creation scientists still try to use the atheistically-designed geologic timeline as an indicator for sediment age during the year-long Flood. The author believes this endeavor is futile and not verifiable because the foundational ideas it is based on are invalid. The Flood was an event so cataclysmic and so far outside the uniformitarian paradigm that it has been impossible, so far, to develop any detailed biblical creationist Flood model consensus.

Ammonite *Mortoniceras*

Figure 55: *Mortoniceras* ammonite fossil from Texas [MARF24].

---

* See Mitchell, J.D., *Discovering the Animals of Ancient Oregon*, Leafcutter Press, 2013, pp. 96-100.

Description: *Mortoniceras* partial fossil is 3" wide × 2 ½" high × 1" thick. The rib pattern is irregular and rib peak spacing is about ¼". Fossil has evolute, open-coiling style and is totally permineralized.

Interpretation: Relevant "18 Facts" > 1, 2, 3, 5. The author purchased this as a typical example of the numerous similar fossils available for sale in Texas at a Glen Rose, Texas rock shop. The owner of the shop related that he had found this type and similar ammonites in numerous counties of Texas, including Florence, Summerville and Hood. The Texas ammonites in his rock shop came in sizes up to three feet in diameter.

This is an example of a fossil that does not show the complete shell and is very rough with little eye appeal. Perhaps this is a result of the mineral composition of the matrix in the areas of Texas where these ammonites are found, the skill of the collector/preparer, or a combination of these things.

Ammonite *Scaphites*

Description: This *Scaphites* ammonite is 1 ⅛" wide × 1" high × ¼"

Figure 56: Fossil ammonite *Scaphites* from the Fox Hills formation of South Dakota [MARF10].

thick. The matrix nodule segment is about 3" diameter × 1 ¼" thick. There are at least a dozen other smaller fossils of different kinds of marine animals visible in the matrix.

Interpretation: Relevant "18 Facts" > 1, 2, 3, 4, 5, 14. The *Scaphites* ammonites are characterized by a J-shaped body design where the chambered early whorls are tightly coiled like most ammonites, but the living chamber does not follow the usual coiling.

The fossil matrix here is part of a nodule that was broken apart to reveal a variety of sea creatures along with the ammonite. The nodule broke along a complex combination of planes that resulted from the locations of the fossils inside.

Ammonite *Douvilleiceras*

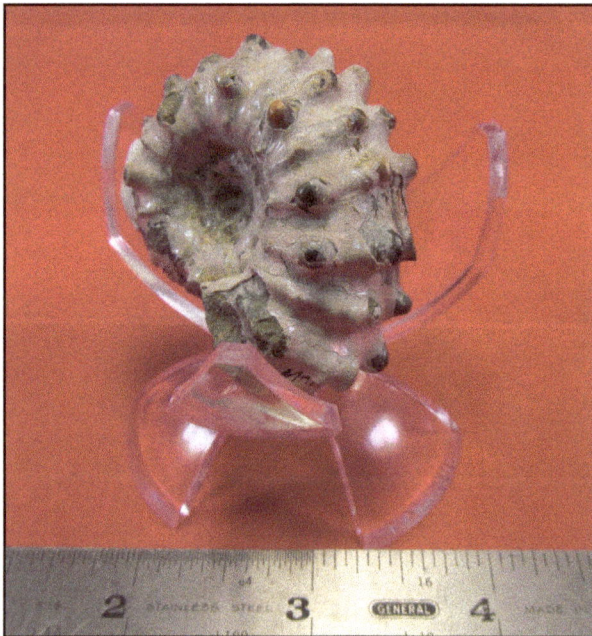

Figure 57: Fossil ammonite *Douvilleiceras* from Madagascar [MARF5].

Description: The *Douvilleiceras* specimen is 1 ¾" wide × 1 ¾" high × 1 $\frac{1}{16}$" thick maximum. The shell is involute and has four rows of tubercles extending from the ribs.

Interpretation: Relevant "18 Facts" > 1, 2, 3, 4, 5, 14, 15. This particular type of ammonite is found in sediments worldwide and is an index fossil for the Cretaceous Period of the atheistic Geologic Timeline.

With a biblical creationist view of the expected actions of the Genesis Flood, it makes little sense that every one of these *Douvilleiceras* fossils would end up at the same sedimentary level around the earth. Index fossils are part and parcel of the evolutionary worldview that the author believes should not be relied on by creation scientists. To do so is to accept the atheistic presuppositions of evolutionary naturalism.

Ammonite *Ancyloceras*

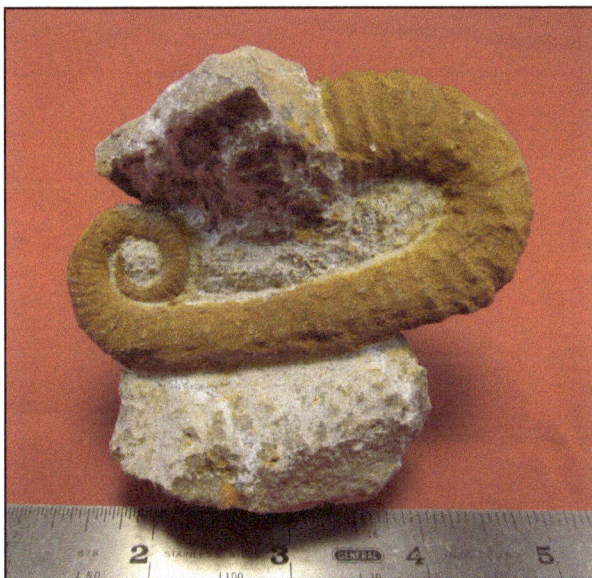

Figure 58: Heteromorph (uncoiled-shell) ammonite fossil *Ancyloceras* from Morocco [MARF58].

Description: Ammonite *Ancyloceras* is 3 $\frac{5}{16}$" long × 1 $\frac{3}{4}$" wide × $\frac{11}{16}$"

thick. The fossil size including the matrix is 3 ¼" × 3" × 1 ¾". The exterior ribs are spaced on ³⁄₁₆" centers.

Interpretation: Relevant "18 Facts" > 1, 5, 6, 7, 8, 9, 10, 14. Evolutionists usually assert that the heteromorphic ammonites are more advanced than the coiled ammonites because the heteromorphs are usually found in higher sedimentary rock layers. Ammonites are popular index fossils and heteromorphic types are used to correlate dinosaur-age sediments for secularists.

The biblical creationist notices that the coiled ammonites actually have a more advanced design, and would have been more streamlined for easier travel in the water. This streamlined capability would have increased their chances of avoiding extinction compared to the heteromorphic-design ammonites. Another way to interpret the stratigraphic order of ammonites and other cephalopod fossils in the rock record would be that the order is determined mostly by where in the water column the animals lived pre-Flood.

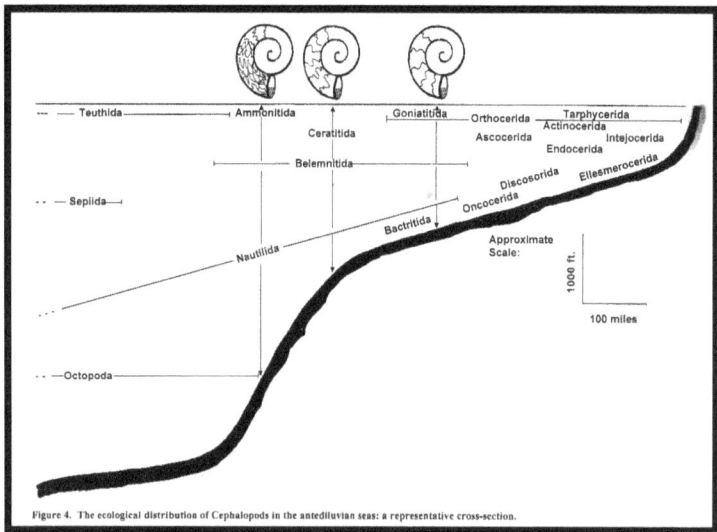

Figure 4. The ecological distribution of Cephalopods in the antediluvian seas: a representative cross-section.

Figure 59: A creationist explanation for the ecological distribution of Cephalopods including ammonites [from Woodmorappe, John, *Studies in Flood Geology*, ICR, 1999, p. 190].

In Figure 59 can be seen how the space in which a particular cephalopod lived pre-Flood could greatly influence where in the rock record the animal would be buried and later discovered. Perhaps *Ancyloceras* lived closer to the surface of the ocean at the time of the Flood than did other ammonite index fossils that are often found lower down in the rock record.

Ammonite aptychi

Figure 60: Fossil ammonite jaws (aptychi) from the Solnhofen limestone of Germany [MARF61].

Description: The aptychi hinge is located precisely in the center of the two halves of the fossil. The jaws are ⅜" wide × ⅜" high. The limestone matrix is 1 ¹¹⁄₁₆" wide × 1 ¾" high × ⅜" thick.

Interpretation: Relevant "18 Facts" > 1, 5, 14. These aptychi "jaws" are only occasionally found associated with a particular ammonite. In those cases they are usually preserved *in situ* close to the ammonite opening. While the most popular explanation by fossil dealers is that the aptychi formed part of a jaw mechanism that was used for shredding food, other ammonite experts believe they were "doors" for closing the opening when the animal's tentacles were retracted.

No matter what the correct explanation, these small plates required very special burial conditions to have been preserved. As with the other complex ammonite parts, the aptychi were carefully designed by the Creator of all things for a particular purpose, or purposes.

Ammonite *Cleoniceras* Ammolite

Figure 61: Fossil ammonite *Cleoniceras* from Madagascar known as an ammolite gemstone [MARF60].

Description: The ammolite is 1 ⅞" wide × 1 ½" high × ¹⁷⁄₃₂" thick. This particular ammonite is only partially infused with minerals, but red, green and yellow colors are visible.

Interpretation: Relevant "18 Facts" > 1, 2, 3, 14. In Romans 1:20

we are told that the creation is clearly seen and is understood by the things that are made. Therefore, rejecters of the Creator God are without excuse!

Beautiful rocks and minerals of almost every imaginable description have been discovered in the crust of the earth. Ammolites are just one of these marvelous natural finds. We can assume that the omniscient Creator God was well aware of this beauty and provided it for our enjoyment. Even the terrible cataclysm of the worldwide Flood resulted in many wonderful items of beauty like the ammolites. Ammolites were officially given gemstone status by the World Jewellery [sic] Confederation in 1981.

The small *Cleoniceras* ammolite above is just ordinary as a gemstone and not particularly valuable. However, some ammolites, especially larger ones from the Canadian Rockies, are valued at many thousands of dollars each.

Ammonite *Baculites*

Figure 62: *Baculites* fossilized segment from the Pierre Shale of Wyoming [MARF77].

Description: The *Baculites* fossil is 1 ¾" long × 1 ¹⁄₁₆" wide × ⁹⁄₁₆" thick at the large end. The suture lines are very complex and look somewhat like tree leaves. The cross-section is oval in shape.

Interpretation: Relevant "18 Facts" > 1, 5, 13, 14. *Baculites* is an extinct cephalopod that had a cone-shaped long shell except for a small spiral coil at the initial stage. The coil is normally lost from the fossil. Most experts consider *Baculites* to be a heteromorphic form of ammonite. The suture lines are often much more complex than those on coiled ammonites with saddles and lobes intricately folded. "Heteromorphs, such as the straight-shelled Baculites are

Figure 63: Cluster of fossil *Baculites* from the Pierre Shale, Pennington County, South Dakota [Denver Museum of Nature & Science].

shaped in such a way that the head would have hung straight down, and so if they could swim at all it would have been vertically and not horizontally."*

Marks' and Palmer's conclusions regarding the orientation and swimming ability of *Baculites* may be influenced by evolutionary presuppositions and incomplete information about these extinct animals. The location in the rock record for *Baculites* is reported

---

* Marks, Neale & Palmer, Philip, *Ammonites*, Smithsonian Institution Press, 2002, p. 93.

to usually be higher than for most other ammonites. The "other" ammonites are considered to be good swimmers by experts. The author asks: How could the loss of swimming ability be considered an evolutionary advantage?

Scaphopod *Dentalium*

Figure 64: Fossil scaphopod *Dentalium* section from Washington State [MARF63].

Description: The fossil shell is $^{11}/_{32}$" outside diameter on the large end and tapers to $^{7}/_{32}$" outside diameter on the small end. The wall thickness is about $^{1}/_{16}$" and the broken section shown is 1 $^{7}/_{16}$" long. There are about 60 fine longitudinal ribs around the circumference of the tube.

Interpretation: Relevant "18 Facts" > 5, 10, 11, 12, 14, 17. The *Dentalium* scaphopod animals are living fossils found worldwide as fossils and as loose "tusk shells." In life the shell is oriented with the large diameter pointing down into the sediment where the animal's foot, tentacles and mouth are located. The digestive track and anus are located inside the tube. The tubular shell starts out quite curved but straightens as the animal grows.

A first glance at a *Dentalium* fossil reveals nothing but a curved tube. Without the living animal that exists on the sea floor today, it is doubtful that anyone would be able to determine that the *Dentalium* fossils represent a once-living complex lifeform. As with all life, the biblical creationist sees design rather than evolution and deep time in these marine animals.

Nautiloid *Orthoceras*

Figure 65: Fossil nautiloid *Orthoceras* from the Sahara Desert in Northern Morocco [MARF13].

Description: Main center specimen of *Orthoceras* is ½" diameter × 4" long and eight septa are visible. The partial specimen at the top edge represents an animal with a shell that was at least ¾" in diameter. The matrix is 2 ⅛" wide × 5 ¾" long × 1 ⅛" thick. A number of other nautiloid fragments of various sizes and orientations are also visible in the matrix.

Interpretation: Relevant "18 Facts" > 1, 2, 3, 4, 5, 14. The *Orthoceras* and other similar fossil nautiloid shells were shaped like a tapering cylindrical cone and had numerous concavo-convex chambers like today's chambered nautilus. They have never been found in today's oceans and are thought to be extinct. Just like spiral nautiloids, they must have had heads with eyes and grasping tentacles. *Orthoceras* are found in massive fossil graveyards in Morocco and workers there shape and polish them into all sorts of artistic items for sale. Because of their beauty, they are very commonly found for sale in rock and fossil, jewelry, and other specialty shops all over the world. It is amazing that fossilized sea creatures such as *Orthoceras* are found in the Sahara Desert. This fact indicates that the area has not always been dry and nearly covered with sand as it now is. Those facts mesh with the biblical creationist expectation of massive changes to the earth's surface being a result of the Genesis Flood.

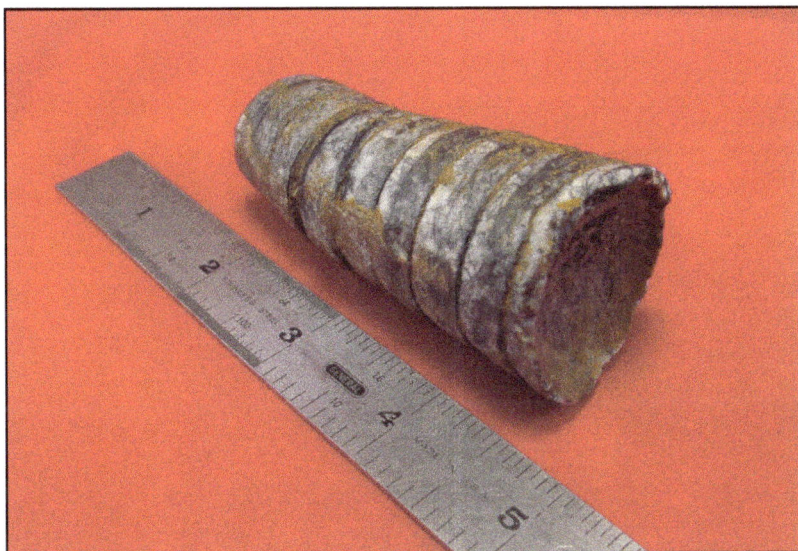

Figure 66: Fossilized section of the nautiloid *Orthoceras* from Morocco [MARF39].

Description: The *Orthoceras* section is 1 ⅜" diameter on the large end × ¹⁵⁄₁₆" diameter on the small end × 3 ⅛" long. There are 8 ½ chamber sections in the fossil. On the large concave end can be seen remnants of the siphuncle that ran up the center of the animal. The shell looks to be made up of uniformly spaced conca-vo-convex chambers.

Interpretation: Relevant "18 Facts" > 1, 2, 3, 4, 5, 14. Since *Ortho-ceras* is extinct, any assumptions about how they lived must be made on incomplete evidence. It can be assumed that the animals that lived in the shells (the phragmocones) were likely similar in design, appearance, and complexity to the nautilus of today. Since the *Orthoceras* fossils are cone shaped we can guess that the animal lived in the large end and swam by ejecting water out that end. That indicates the animal probably moved mostly horizontally in the direction of the small end. The Moroccan-found *Orthoceras* fossils are usually a maximum length of about six inches. The loose out-of-matrix *Orthoceras* specimens like the one in the figure, tend to break apart at the chamber joints so that complete fossilized animal shells are rare.

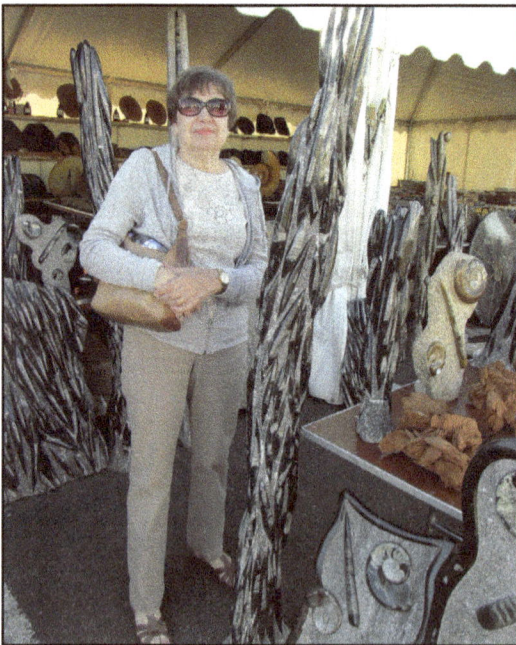

Figure 67: Loads of *Orthoceras* nautiloid fossils as revealed in the limestone of Morocco [Denver Fossil Show].

Description: The artistic slabs consisting of hundreds of *Orthoceras* nautiloids behind and to the left and right of the lady in Figure 67 are up to seven feet long. The nautiloids are consistently oriented parallel to each other indicating a common direction of travel when they were rapidly buried.

Interpretation: Relevant "18 Facts" > 1, 2, 3, 4, 5, 16. How can the usual orientation and condition of these fossil nautiloids by the billions be logically explained using secular presuppositions? The answer is that they cannot!

In the Grand Canyon, "billions of large fossilized orthocone nautiloids are entombed in a six foot thick layer near the base of the massive Redwall Limestone formation. This extraordinary layer persists throughout the Grand Canyon region, northern Arizona and southern Nevada. The fossil bed occupies an area of at least 5,700 square miles and contains an average of one fossilized nautiloid per square yard."*

The evidence provided by these nautiloids in Morocco and the Grand Canyon area screams of catastrophe unlike any of the natural events we see today. The global Genesis Flood is the only logical explanation for the *Orthoceras* massive graveyards!

Belemnite Guards

Description: Guard segment (Figure 68) at the left is 2 ½" long × ½" diameter maximum. Right guard is 2 ¼"long × ½" in diameter. The right hand specimen exhibits a furrow along one side that is characteristic of many belemnite fossils.

Interpretation: Relevant "18 Facts" > 1, 2, 3, 4, 5, 14. Belemnites are now-extinct cephalopods similar in design to the living squid, cuttlefish and octopus. They had bullet-like internal skeletons called guards that are the usual parts found fossilized in

---

* Vail, Tom, *Grand Canyon – a different view*, Master Books, 2003, p. 52.

Figure 68: Two fossil end segments of the guards of an unknown genus of belemnite (might be *Belemnitella*) from the Black Ven marl of Lyme Regis, Dorset, U.K. [MARF46].

sedimentary rock worldwide. It is thought that the guards served as ballast for the animals.

Squids possess an internal support called a pen made of chitin. In belemnites the internal support (guard) is made of calcium carbonate. Both animals were wonderfully designed by the Engineer of the universe. Perhaps belemnites went extinct because they were unable to survive the conditions created by the cataclysmic Genesis Flood.

Nautilus *Cenoceras*

Description: The sliced *Cenoceras* fossil MARF18 (Figure 69) is 3 ⅝" wide × 3" high × 1 ⅛" thick at maximum. There are 38 countable concavo-convex chambers connected by a siphuncle running through the center of the chambers. The largest (last) chamber is ⅜" wide × 1 ⁷⁄₁₆" long. The last living chamber was broken away and is not present in this fossil. The whole *Cenoceras* fossil MARF14

Figure 69: Cut and polished *Cenoceras* nautilus fossil (left) and polished whole *Cenoceras* nautilus fossil (right). Both fossils are from Madagascar [MARF18 - MARF14].

Figure 70: Extant chambered nautilus shells - sliced shell (left and whole shell (right).

is 3 ¾" wide × 3" high × 2 ⅝" thick. This fossil is very similar in size to the sliced specimen.

The extant cut and whole chambered nautilus shells shown in Figure 70 are about the same size (4 ⅛" wide × 3 ⅛" high × 2 ⅛" thick). There are 23 countable concavo-convex chambers in the cut specimen. The extant shells have the same involute shell spiral form as do the fossil specimens.

Interpretation: Relevant "18 Facts" > 1, 5, 14, 15, 17, 18. Since the terminal living chamber of the cut fossil is broken away it represents a living animal quite a bit larger than the two extant shells. There is no significant difference in design between the *Cenoceras* and the chambered nautilus. Therefore, *Cenoceras* is another example of a living fossil, and these specimens were rapidly buried in the Genesis Flood some 4,500 years ago.

Figure 71: Another chambered nautilus shell sliced so that siphuncle path can be seen.

Starfish *Stenaster*

Description: The fossil starfish fits inside a 3 ¼" diameter circle and is up to ⅛" thick. Fine exterior original markings are visible on three of the five arms. Also visible in the matrix are partial bodies of three trilobites. Matrix is 5" × 5 ½" × 1 ½" thick. The recent starfish body fits inside a 5 ¾" diameter circle and is ¾" thick.

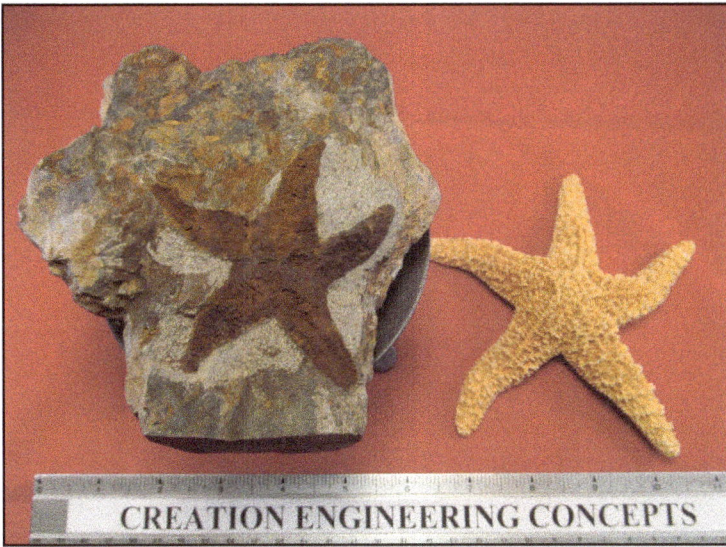

Figure 72: *Stenaster* starfish fossil from Morocco (left) [MARF35] and common (sugar) starfish from Oregon shell shop (right).

Interpretation: Relevant "18 Facts" > 1, 5, 9, 10, 17. Experts have identified scores of species names for fossil and extant starfish (Asteroids) and oftentimes do not agree on the proper genus/species for them. Asteroids are often defined as having five arms broadly connected to the central body and looking like a five-pointed star. Asteroid invertebrate animals have a mouth that is located on the underside in the middle. These descriptions match both starfish in the photo.

The author displayed these two starfish in the photo next to one another to highlight the close similarity in shape and design. If they should be the same in other aspects of construction then *Stenaster* and the common (or sugar) starfish could be the same genus and be another living fossil.

Brittle Star *Geocoma*

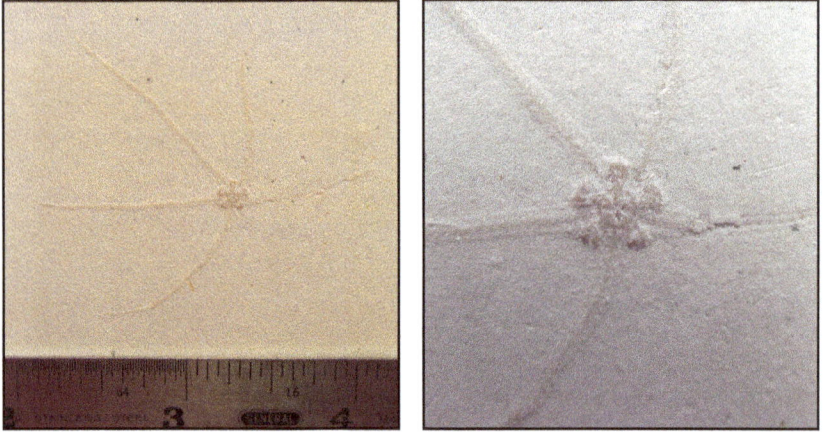

Figure 73: Fossil brittle star *Geocoma* from the Solnhofen limestone of Bavaria, Germany [MARF3].

Description: The *Geocoma* center disc is ³⁄₁₆" in diameter and the five fragile arms extend to form a circle about 2" in diameter. Under magnification the segments of the arms can be clearly seen. The rightmost arm has sections missing that were lost during recovery of the fossil when the slab was split open. The limestone matrix sheet is 5 ¼" wide × 4" high × ⅛" thick.

Interpretation: Relevant "18 Facts" > 1, 5, 7, 14, 15, 17. The author purchased this fossil because it was recovered from the same type of fine limestone in which several of the famous *Archaeopteryx* specimens were found. The matrix is so fine that amazing detail in fossils can be seen such as the detail of construction of this very fragile brittle star.

The arms and center disc of brittle stars (Ophiuroids) are much different from those of starfish (Asteroids). The *Geocoma* fossil brittle star seen here is very similar to some existing brittle stars that inhabit the ocean floors in vast masses today. When fossil brittle stars are recovered they too are often found in vast quantities.

Echinoids *Pygurus* & *Encope*

Figure 74: Sand dollars: left one is recent and was purchased from an Oregon beach gift shop. Right one (*Pygurus* or *Clypeaster*) is fossilized and is from Madagascar [MARF38].

Description: The non-fossilized sand dollar is 2 ⅝" diameter × ¼" thick. The fossil *Pygurus* or *Clypeaster* sand dollar is 2 ⅞" diameter × ⅞" thick and has been polished as are most fossils from Madagascar. Both specimens have five "petals" on the upper surface.

Interpretation: Relevant "18 Facts" > 1, 4, 5, 17. Sand dollars are similar to sea urchins and are classified as echinoids. The rigid globular skeleton is the "test" or shell and the animal has its mouth located on the underside at its center. Inside the mouth are five small teeth (called "doves") that break down the plankton diet of the animals. The petals on the top are used for respiration. Sand dollars are found worldwide and have been assigned many different species names.

The biblical creationist concept of "variation within created kinds" explains the fact that sand dollar species are found on beaches today that look very much like those in the rock record. Sand dollars are examples of living fossils.

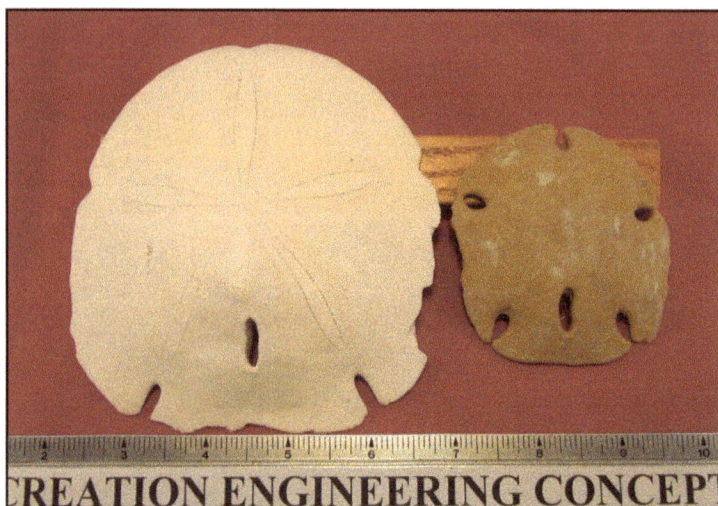

Figure 75: Left sand dollar is recent and was purchased from an Oregon beach gift shop. Right one, *Encope*, is fossilized and is from an unknown location in North America [MARF41].

Description: Both of these sand dollars look to be the same genus *Encope*. Larger specimen is 5" diameter × ¾" thick and has three oval holes visible and three edge "inlets." The fossil specimen is 3" in diameter × ⁷⁄₁₆" thick and has six oval holes in its test (shell). Both animals had five petals on the upper surface that they used for respiration.

Interpretation: Relevant "18 Facts" > 1, 4, 5, 17. When a living sand dollar is found the outside surface of its test is covered with millions of tiny spines that aid in the movement and feeding of the animal. The oval-shaped holes in the test of *Encope* are also thought by evolutionists to have been developed by the animal itself over millions of years to aid in feeding.

Encope and *Clypeaster* sand dollars are very similar except for the oval holes that pass through the test on *Encope*. Could it be that this modification was provided by God simply for variety in appearance? At any rate, both *Encope* and *Clypeaster* are additional examples of the hundreds of living fossils found so far that speak against the reality of macroevolution.

Coral *Heterophrentis*

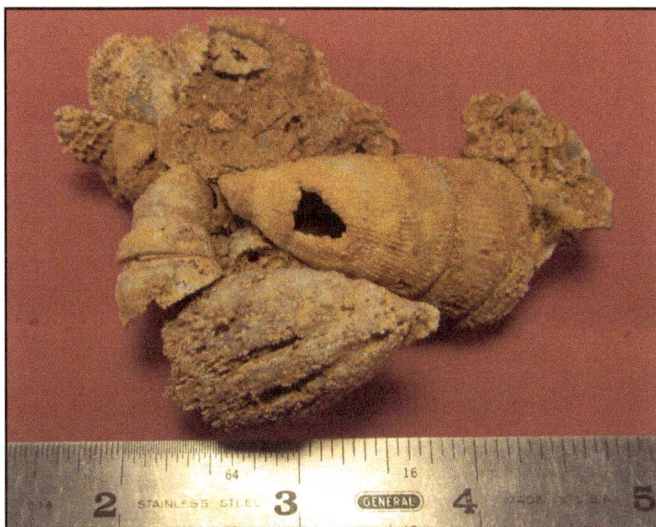

Figure 76: A conglomerate of four fossil *Heterophrentis* horn coral from the Jeffersonville limestone of Indiana [MARF45].

Description: These *Heterophrentis* animals in life were cone-shaped and lived a solitary life. On the exterior they had long longitudinal ridges and encircling growth lines. The largest horn coral specimen in the figure is 1" diameter × 2" long. The complete fossil conglomerate is 2" wide × 3" long × 1" thick. The matrix consists largely of numerous coral fragments and other unidentifiable matter.

Interpretation: Relevant "18 Facts" > 2, 3, 5, 12, 14, 16. Secular scientists describe coral as "simple" lifeforms. Yet they are complex beyond man's ability to duplicate or create them from nothing. Fact #12 states that "all scientists observe the same evidence but have varying interpretations depending on their presuppositions." A case in point is the rate of growth of today's coral reefs. Uniformitarianism dictates that reef growth is too slow to have allowed their current size within the biblical timeline. Creation science rejects uniformitarian presuppositions and accepts that events could have transpired at much faster rates in the unobservable past.

Coral *Astrhelia*

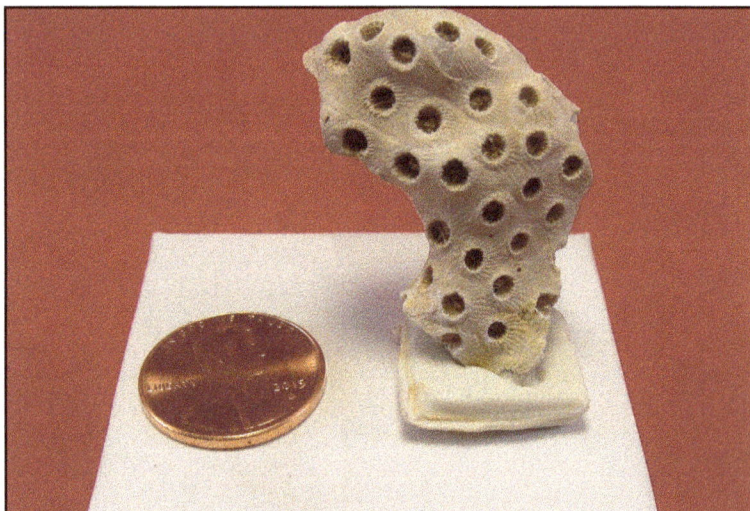

Figure 77: Fossil coral *Astrhelia* from Calvert County in Maryland [MARF59].

Description: The *Astrhelia* coral has exterior small circular calices nearly evenly spaced. The calices are $\frac{1}{16}$" to $\frac{1}{8}$" diameter with approximately thirty calices per square inch. The fossil piece is 1 $\frac{1}{2}$" tall × $\frac{3}{4}$" wide × $\frac{1}{2}$" thick. The surface, septa and calices are all sharp with very little water wear.

Interpretation: Relevant "18 Facts" > 1, 5, 10, 11, 12, 14. This particular genus of coral is categorized as a recent evolutionary development by evolutionists. But from what did it evolve? This question is a common problem in evolutionary theory for many kinds of plants and animals. These corals and similar kinds have not been found deep in the sediments and so the evolutionists must assume they have an unknown ancestor.

    When the biblical creationist looks at this specimen he sees tremendous design and much microscopic detail. These attributes do not come from the random-chance interaction of matter. They do result from the intentional application of knowledge by an intelligent mind. We know men did not create this coral design so it makes sense that God did, as we are told in the Bible! The biblical creationist agrees that there is variation within created kinds, but this variation is always limited.

Coral Colony *Phillipsastrea*

Figure 78: Fossilized hemispherical colony of *Phillipsastrea* coral from Morocco [MARF49].

Description: The corallites on the surface of the coral colony are round to irregularly-shaped polygons and vary in size from $\frac{1}{8}$" to $\frac{1}{4}$." There are about six corallites per square inch of surface. The *Phillipsastrea* coral colony is 4" × 4 $\frac{1}{2}$" × 1 $\frac{1}{2}$" thick.

Interpretation: Relevant "18 Facts" > 1, 2, 3, 4, 5, 12, 15, 17. According to secular thought, these fossilized colonies of coral are millions of years old. Yet, similar coral colonies exist today. Why don't they fossilize by the billions? The reason is that fossilization almost always requires the special conditions of rapid burial under large amounts of sediment-filled water. Those are the unique conditions of the Genesis Flood, a one-time-in-history event.

Coral Colony "Chrysanthemum"

Figure 79: Rough-cut specimen of "Chrysanthemum" fossil coral from Taiwan [MARF50].

Description: The corallites are about ³/₁₆" diameter with about 25 corallites per square inch. The slab is 2 ½" × 2 ½" × ⁷/₁₆" thick.

Interpretation: Relevant "18 Facts" > 1, 2, 3, 5, 14, 17. Various genera of fossilized colony-type coral from all over the world are sliced, cut and polished into beautiful cabochon pendants and jewelry. Besides Chrysanthemum seen here, another popular coral stone for these purposes are the Petoskey stones (*Hexagonaria*) from Michigan.

How amazing is the natural beauty of millions of created things God has provided, including fossils from the worldwide Flood!

Dragonfly *Protolindenia*

Figure 80: Replica fossil dragonfly *Protolindenia* with 6 ½" wingspan from the Solnhofen limestone in Germany [OR1].

Description: This large dragonfly's body was about 3 ¼" long × ¼" wide. The matrix size is 4 ½" wide × 7" long.

Interpretation: Relevant "18 Facts" > 6, 14, 15, 17. Large fossilized dragonflies have been found in sedimentary rock throughout the world including a fairly large number in the Solnhofen limestone of Bavaria. Some fossil dragonflies have been reported to have had wingspans of up to three feet. These large insects probably existed pre-Flood because of the considerably different climatic and environmental conditions at that time. Some secular and creation scientists are in agreement that one likely condition in the past was an atmosphere with a higher concentration of oxygen than we now experience. Other than size, there is little difference between the fossil and existing dragonflies. All dragonflies, whether fossil or living, are extremely complex lifeforms with amazing flying capabilities.

Mayfly Larvae *Ephemeropsis*

Figure 81: Fossil mayfly larvae (*Ephemeropsis*) in matrix from the Yixian formation, Liaoning, China [MARF40].

Description: The *Ephemeropsis* fossil is 2 ⅜" long × ⅜" wide. The matrix block is 2 ½" wide × 2 ¼" high × 1" thick. The three rear projections from the insect larvae are ⅝" long.

Interpretation: Relevant "18 Facts" > 1, 5, 14, 17. The China formation where this (and other) mayfly larvae (nymphs) were found is similar to limestone formations elsewhere; Solnhofen in Germany, Green River formation in Wyoming and the Lagerstatte in Lebanon are examples. Tiny details can be seen in these fossils that show extreme complexity. No evolution from simple to complex is ever observable in reality.

Those familiar with mayfly larvae of today will not notice much, if any, difference between them and this larvae formed 4,500 years ago and captured in this fossil.

Crane Flies

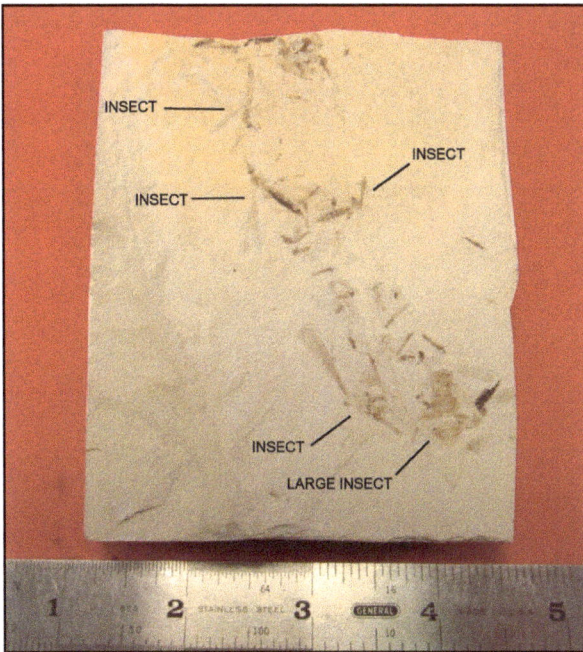

Figure 82: Limestone block with numerous fossil insects from the Green River formation near Bonanza, Utah. The largest insect bodies are indicated [OF6].

Description: The upper four noted insects are probably crane flies as determined by their body and wing shapes. The lower "large" insect type is unknown. The crane fly wings are ⅝" long × ⅛" wide and the bodies are about ⅜" long. The limestone block is 3 ¼" wide × 3 ¾" high × ⅜" to ⅞" thick.

Interpretation: Relevant "18 Facts" > 1, 2, 3, 4, 15, 17. If delicate insects like these were commonly fossilized by the conditions existing today we should be able to find them in almost every sedimentary rock we pick up, but that is not the case. According to insect experts there are between two and thirty million insect species in existence on the earth, and they make up over eighty per cent of the world's identified specie names.

These delicate insects shown above were surely fossilized by some very unusual processes – probably those of the Genesis Flood at the time of Noah.

Fossil Parasitic Wasp

Figure 83: Fossil parasitic wasp (family Ichneumonid) on matrix from the Green River formation, Rio Blanco County, Colorado [OF30].

Description: Fossil parasitic wasp at the top of matrix is ⁵⁄₁₆" long with wings ⅛" long and antennae ⅛" long. The matrix is 1 ¾" wide × 1 ⅜" high × ⁵⁄₁₆" thick.

Interpretation: Relevant "18 Facts" > 1, 8, 9, 10, 15, 17. The Bible teaches that in the beginning "everything was very good" (Genesis 1:29-30). That leads the biblical creationist to the conclusion that parasitic wasps did not exist in the beginning. After the fall and the institution of the curse by God, many things that were created good changed and now are not so good.

Some zoologists have written that their research indicates that the differences between non-parasitic and parasitic wasps are not as great as might be assumed. "Free-living species could become

parasitic without substantial anatomical or physiological changes."*
In the present cursed world these parasitic wasps still do provide
an important service. They work to control overpopulation of certain insects that the wasps kill as part of their own life cycle.

Beetle *Hydrophilus*

Figure 84: Slab of La Brea Tar Pits matrix with three
embedded *Hydrophilus* fossil water beetles [OF1].

Description: The two *Hydrophilus* beetles at the left are damaged, but
the right hand specimen is complete and is 1 $\frac{1}{8}$" long × $\frac{5}{8}$" wide. The
matrix slab of tar sand is 3 $\frac{1}{2}$" wide × 6 $\frac{1}{2}$" long × 1 $\frac{1}{2}$" thick.

Interpretation: Relevant "18 Facts" > 1, 2, 3, 5, 17. There exists a
high concentration of fossilized water beetles at the La Brea Tar
Pits in Los Angeles, California and that could be interpreted as
an indication that the fossil site is the result of a massive water
catastrophe. The Genesis Flood explanation for the thousands of

---

* Miller, Stephen and Harley, John, *Zoology 8*th *edition*, McGraw-Hill Co., 2009, p. 226.

fossils found at La Brea provides many explanatory advantages over the secular uniformitarian "entrapment" scenario.*

Insects in Amber

Figure 85: Insects trapped in fossil resin (amber) from the Baltic Sea area [OF13].

Description: The piece of fossil resin is 1 $\%_{16}$" long × ¾" wide × $\%_{16}$" thick. There are at least 20 insects trapped in this resin sample that has been artificially polished on its surfaces post discovery.

Interpretation: Relevant "18 Facts" > 1, 12, 17. Resin samples that are called amber or copal are collected for resale from Columbia and the Dominican Republic as well as from the Baltic Sea areas of Lithuania, Latvia, Estonia, Poland, and Russia. Since trees today make resin and the resin can at any time trap insects and harden, there is no scientific method generally used to prove the millions

---

*   See Mitchell, J.D., *The Creation Dialogues – 2ⁿᵈ Edition*, CEC Publications, 2014, pp 76-83.

of years that the amber is advertised to be. If radiocarbon (carbon 14) dating were used along with the proper biblical assumptions about the age of the earth, the likely age of all hardened "fossil" resin would be less than 4,500 years. And, the author believes that the Baltic fossil resin in the photo is actually about 4,500 years old and was formed in the worldwide Flood.

The author has inspected hundreds of photos of fossil resin with trapped insects and other lifeforms inside and has yet to find an example where the entrapped life looked any different from the life kinds living today. It was not surprising that these photos showed no hint of macroevolution over the age of the resin.

Science is no closer to making a dinosaur or any other animal from the DNA in the blood inside a mosquito in amber than it was at the time of the first *Jurassic Park* movie. Real science supports the Bible's instruction regarding created kinds at the beginning and does not support the atheistic evolutionary fairy tales promoted by our secular culture.

Trilobite *Paralejuris*

Description: The *Paralejuris* fossil trilobite in Figure 86 is 3" long × 1 ¼" wide × ⁹⁄₁₆" thick. The matrix is 3 ½" wide × 4 ⅛" long × 1" thick. The head shield is convex and nearly semicircular. There are ten segments of the thorax and the two eyes are positioned at the rear of the glabella (head shield).

Interpretation: Relevant "18 Facts" > 1, 2, 3, 5, 8, 14. Trilobites have three parallel sections or lobes that make up their bodies; and thus the name *tri-lobe-ite*. These arthropods are found in the rock record by the billions all over the earth. They have many variations and many genus and species names have been assigned to them. They range in size from less than ¹⁄₁₆" long to over a yard long. They are thought to be extinct and are often described by collectors as "bugs." Most experts believe that they are most closely related to the horseshoe crab of today.

Figure 86: Fossil trilobite *Paralejuris* from Morocco in matrix [MARF9].

Paralejuris is a genus among dozens found in Morocco that local entrepreneurs extract, prepare and then sell all over the world. Some of these trilobites are crudely extracted from the rock matrix using a hammer and a nail and others are prepared using the latest paleontological tools and techniques. The particular specimen shown above was prepared with a technique sufficiently advanced to show the complexity of the two eyes and the series of pleural segments along the sides of the animal.

SOME TRILOBITE TERMINOLOGY

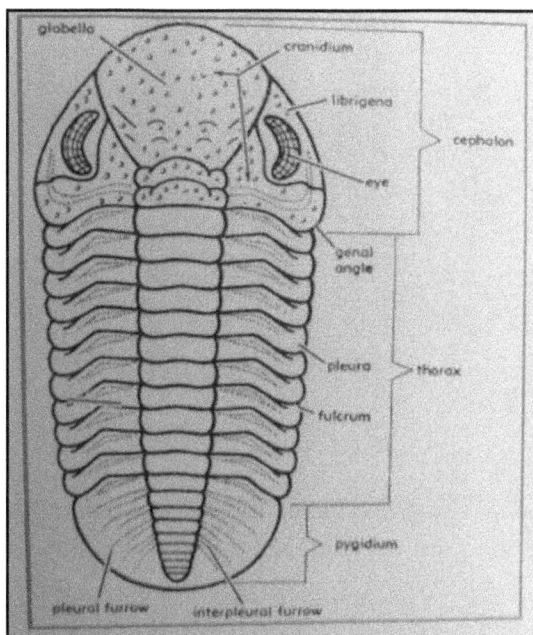

Figure 87: Trilobite Anatomical Terminology.

Trilobite *Calymene*

Description: The *Calymene* trilobite (Figure 88) in the sandstone matrix is 1 ¾" long × ⅞" wide × ½" thick. The matrix is 1 ¾" long × 1 ⁹⁄₁₆" wide × ¾" thick. The loose specimen is 3" long × 3" wide × 1" thick. Detail of the smaller specimen allows for seeing hundreds of lenses in the two eyes and the 13 grooved thoracic segments. In the larger specimen only ten of the thoracic segments are visible and the eyes are obliterated.

Interpretation: Relevant "18 Facts" > 1, 2, 3, 5, 8, 14. The design of trilobites with a tough external exoskeleton permits them to be found fossilized in the rock record by the millions. Why trilobites became extinct cannot be known for sure, but they demonstrate God's marvelous design and the foolishness of evolutionary thought.

Figure 88: Fossil trilobites *Calymene* from Morocco [MARF42 & MARF8].

Many trilobite genera had compound eyes with hundreds of lenses composed of inorganic, crystalline calcium carbonate (calcite). Unlike most animal eyes that are made of organic tissues that are not preserved as fossils, the lenses of trilobites are often perfectly preserved. It is also significant that some of these trilobite lenses are double lenses to allow sight under water without distortion.

Trilobite *Elrathia*

Description: The smaller *Elrathia* trilobite in Figure 89 is ⅜" long × ⁵⁄₁₆"wide in a matrix 2 ¼" × 2 ¹⁄₁₆" × ⁵⁄₁₆" thick. The larger *Elrathia* is 1 ³⁄₁₆" long × ¾" wide in a matrix 4" × 2 ½" × ¾" thick. It has 17 segments in its thorax and is missing the right-hand portion of its head shield. To the left of the larger trilobite is a portion of the front end of another *Elrathia* trilobite. The small eyes of the trilobites are not discernable in these specimens.

Interpretation: Relevant "18 Facts" > 1, 2, 3, 5. There are so many

Figure 89: Two fossil trilobites *Elrathia* in separate pieces of Wheeler Shale matrix from Millard County, Utah [MARF44 & MARF47].

of these particular *Elrathia* trilobites found in the shale rocks of Millard County, Utah that this genus has been designated as the state fossil of Utah. Mass mortality plates with many trilobites together in a "graveyard" are quite common, and is a strong indication of the actions of the Genesis Flood that buried and allowed for the fossilization of these fascinating animals.

Trilobite *Acadagnostus*

Description: *Acadagnostus* in Figure 90 is only ⁵⁄₁₆" long × ⅛" wide in a matrix 1 ¾" × 1 ⅛" × ⅜" thick.

Interpretation: Relevant "18 Facts" > 1, 2, 3, 5, 14. The *Acadagnostus* trilobites are called "blind" trilobites because there are no indications that they have eyes. In Utah they are often found in the same shale sediments as *Elrathia* and are usually very small like the one in the figure.

Figure 90: Fossil trilobite *Acadagnostus* in the Wheeler Shale matrix from Millard County, Utah [MARF48].

Trilobite *Phacops*

Figure 91: Enrolled fossil trilobite *Phacops* on matrix from Morocco [MARF52].

Description: The *Phacops* trilobite is ¾" wide × ⁹⁄₁₆" thick × 1 ³⁄₁₆" long. It has 12 grooved thoracic segments and the multifaceted eyes are clearly visible. The specimen with matrix is 1" × ¹³⁄₁₆" × 1 ⅝" long.

Interpretation: Relevant "18 Facts" > 1, 2, 3, 5, 14. The *Phacops* genus is found fossilized worldwide. While other genera of trilobites are sometimes found enrolled, *Phacops* seems to be found that way more than any other genus. For these animals to be found so often in the enrolled orientation is good evidence that they were quickly "frozen" that way by large amounts of water-laden sediment. This is just what the biblical creationist would propose is a result of the global Genesis Flood.

The best guess is that the enrolled position was a defensive position for the trilobites.

Trilobites *Crotalocephalus* and *Reedops*

Description: *Crotalocephalus* is 2 ¾" long × 1 ⅛" wide × ½" thick. It has a distinctive thoracic design with the center axis dominating the outside pleural lobes. There are 14 thoracic segments and an additional 2 ½ segments that are part of the cephalon. The glabella is bulbous and makes up ⅜" of the overall trilobite length. The eyes are relatively small and located low on the cephalon. The pygidium has four trailing spines.

Reedops is 1 ⅜" long × ¾" wide × ⅜" thick. Its multi-faceted eyes are prominent high on the cephalon and it has about a dozen thoracic segments. It has many characteristics of *Phacops* and it is difficult to tell the difference between the two genera. The main difference is that *Reedops* has a more bulbous glabella (Figure 87) than does *Phacops*.

The matrix is 3 ¼" long × 3 ½" wide × 2" thick including the trilobite.

Interpretation: Relevant "18 Facts" > 1, 2, 3, 4, 5, 8, 9, 12, 13, 14, 18. Secular scientists promote the idea that trilobites came on the scene from an unknown ancestor 524 million years ago, evolved

Figure 92: Sandstone block with two trilobites: *Crotalocephalus* (left) and *Reedops* (right) from Alnif in southeast Morocco [MARF89].

into many different species over a period of 273 million years, and then went extinct. However, the evolutionary family trees for trilobites only show complete, fully designed and functional trilobite kinds with no kind clearly in the process of morphing into another kind of organism. To the author the various evolutionary trees look like made-up fairy tales.

The two trilobites in the figure above have many similarities, but the differences are major. Do evolutionary hypotheses explain the relationship between these two animals that were instantly buried at the same time and place?

The author believes that the creationist explanation is superior. Trilobites of several kinds were created on Day Five of the creation as explained in the Bible. After the fall, and for a period of 1,600 years until the Flood, variations of trilobites were developed and then buried in the Flood. Like numerous other kinds of life, they were not able to re-establish after the Flood and became extinct. Thousands of years after the Flood the fossils were found, excavated, and then prepared for display and study.

Trilobite *Cambropallas*

Figure 93: *Cambropallas* fossil trilobite from the Atlas Mountains of Morocco in Africa. Mold (left) and cast (right) [MARF82].

Description: The *Cambropallas* fossil trilobite is 9" long × 5 ⅝" wide in a nodule that is 10 ½" long × 6 ⅞" wide × 3" thick. The trilobite has 17 thoracic segments. The genal spines are pressed tightly into the thorax on each side.

Interpretation: Relevant "18 Facts" > 1, 2, 3, 4, 5. This specimen exhibits signs of having been compressed not only vertically, but also laterally. The front cephalon portion of the *Cambropallas* extends toward the back with genal spines on both sides. It can be assumed that these genal spines may have assisted in the swimming capabilities of the trilobite. The orange color is due to chemical coating by iron in the sediments that worked to fossilize the trilobite after it was rapidly buried in the global Flood about 4,500 years ago.

Cambropallas is one of the trilobite genera most often faked by Moroccan fossil dealers. The specimen above consists of both a mold and a cast in a nodule so it is very certain that it is authentic. Some dealers like to describe this type of fossil as a trilobite in a "coffin." The large non-authentic Moroccan trilobite fossils are usually made from the cast of a real fossil and are cleverly cemented onto a slab of rock with numerous tool marks oriented around the trilobite to make it look like it was actually carved out of the rock.

Figure 94: All excepting for one of these ten large Moroccan trilobite fossils are not authentic. Can you determine which one is authentic? (Ans.: Rear one oriented horizontally.)

Trilobite *Dicranurus*

Figure 95: Fossil replica of the *Dicranurus* trilobite that has been found in Oklahoma and Morocco [OR26].

Description: The replica trilobite is 3 ¼" wide × 3 ¾" long × ¾" thick. The eyes protrude above the cephalon and it has ten segments along its thorax that extend outside its body as spines of varying lengths. Two additional spines extend sideways from the cephalon that are about 1" long. The largest spines are 2 ¼" long and the body of the trilobite less the spines is 1 ½" wide × 2 ¼" long. There are two curled spines that extend from the posterior of the cephalon. If these spines were extended to be straight they would be about 1 ¼" long. The matrix is 3 ½" wide × 5" long × 1" thick.

Interpretation: Relevant "18 Facts" > 8, 9, 10, 14, 18. As seen in this specimen, trilobites have been discovered with tremendous variation in design and exterior ornamentation. The marvelous spines on *Dicranurus* indicate design for a purpose. Could the purpose be defense, offense, attraction or something else? The name *Dicranurus* means "twin head-tail," an appropriate reference to the two curved spines that it displays.

It is believed that all trilobites had appendages that extended downward to allow them to walk or swim and to breathe underwater. The two-part legs had a locomotion section and also had a separate structure that acted as a gill. Each thoracic segment bore one pair of legs and the cephalon and the pygidium also connected to legs in a similar manner. Since these soft structures were very seldom fossilized, paleontologists continue to develop hypotheses for the details, structure, and physiology of the legs. See Figure 232 for a photo of a trace fossil showing trilobite tracks.

Trilobite *Ellipsocephalus*

Figure 96: Plate with multiple fossil *Ellipsocephalus* trilobites from the Jince formation in the Czech Republic [MARF74].

Description: The *Ellipsocephalus* trilobites are ¾" to 1" long × ½" to ⅝" wide. The matrix is 5" wide × 3 ½" high × 1" thick. There are four complete trilobite bodies, two partial trilobites and four casts.

Interpretation: Relevant "18 Facts" > 1, 2, 3, 5, 15. As mentioned earlier, trilobites are found as fossils by the billions and are often found with many bodies and/or molts together or piled on top of each other in fossil graveyards like the one in the photo. There are ten different *Ellipsocephalus* fossil trilobites represented on this plate, and this is another example of rapid burial due to the actions of the global Flood.

Trilobite *Eldredgeia*

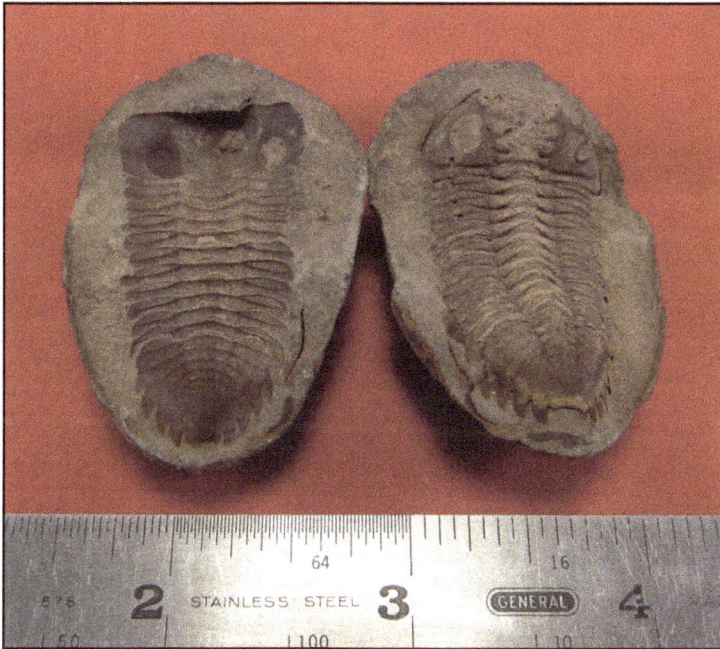

Figure 97: Fossil trilobite *Eldredgeia* in a sandstone nodule with positive and negative molds from the Belen formation in the Altiplano near La Paz, Bolivia [MARF81].

Description: This trilobite is 1 ½" long × ⅞" wide and the encasing nodule is 2" long × 1 ¼" wide × 1" thick. The right eye is intact but part of the left eye is stuck to the negative mold. The thorax has 21 segments and the tail shield has four spines extending from the pygidium toward the rear on each side.

Interpretation: Relevant "18 Facts" > 1, 4, 5, 14, 15. *Eldredgeia* is the most common trilobite found in South America and is usually found in nodules like the one in the figure above. The city of La Paz in Bolivia is at an elevation of 10,500 to 13,500 feet and so this trilobite was recovered from the top of a mountain! This can best be understood from the Bible's explanation that the waters of the Genesis Flood rose to be higher than the highest pre-Flood mountains (See Genesis 7:19-20).

Trilobite *Nankinolithus*

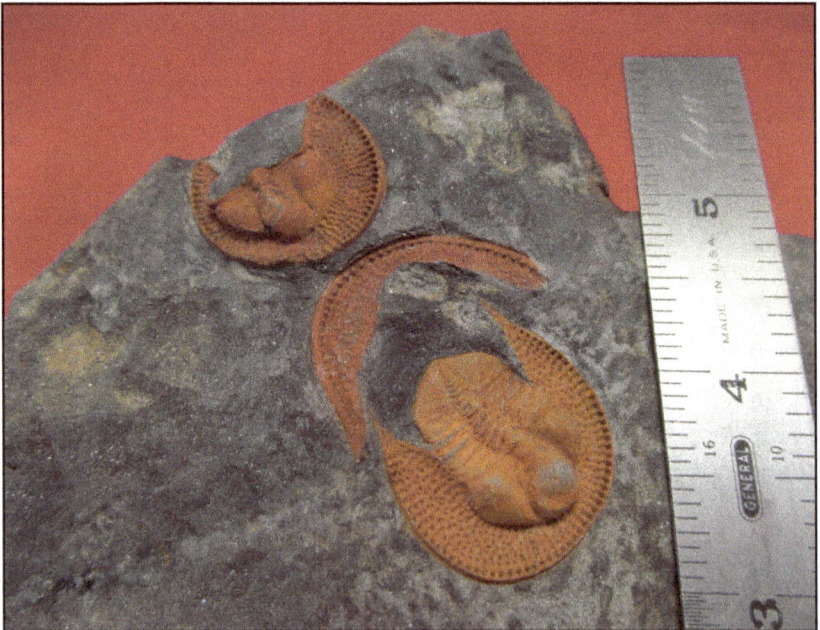

Figure 98: One complete fossil *Nankinolithus* trilobite with parts of two other *Nankinolithus* trilobites in sandstone matrix from Tinjdad, Morocco [MARF85].

Description: The complete *Nankinolithus* trilobite is 1 ¹⁄₁₆" wide × 1 ¹⁄₁₆" long and the matrix block is 4 ½" wide × 7" long × ⅝" thick. The partial (same-sized) fossil trilobite has lost its thorax and pygidium. In between the two fossils is an upside down section of genal spine fringe only. The complete trilobite has 13 or so thoracic segments. There are about sixty perforations in the outside edges of each of the two complete fringes with the total number of interior pits at least four times sixty. The glabellas are highly vaulted and the axis is about one-half the width of the two outside pleurons. The fossils have picked up the pretty orange color from iron in the water.

Interpretation: Relevant "18 Facts" > 1, 2, 3, 5, 8, 9, 10, 14, 18. *Nankinolithus* is part of the marvelous Trinucleidae family of trilobites that all have the unusual perforated headshield or fringe. Since no scientist has seen one of these trilobites alive it is not possible to know with certainty the purpose of the perforated fringe. Perhaps it performed a sensory purpose to keep the trilobite aware of what was in front of him. Some experts have suggested its role was to filter food particles in some manner.

It is undoubtedly true that the fringe had a useful purpose as is the case for every other characteristic of living things. Evolutionists believe that the trilobites themselves, with the help of natural selection, developed the fringe characteristic basically on their own. The biblical creationist would attribute the fringes to God, either directly as created, or through adaptive variation through the God-placed DNA that made them what they were.

Trilobite *Asaphus*

Description: This *Asaphus* trilobite (Figure 99) is 1 ¾" long × 1 ³⁄₁₆" wide on a matrix that is 2 ⅜" × 3" × ⅜" thick. The trilobite has eight thoracic segments. The eye stalks are ³⁄₃₂" diameter × ¼" long and the eye bulbs are ⅛" × ⅛".

Interpretation: Relevant "18 Facts" > 1, 5, 8, 9, 10, 14. The eye-stalk

Figure 99: Fossil trilobite *Asaphus kowalevskii* in limestone from Vilpovitsy Quarry, St. Petersburg region, Russia [MARF84].

feature suggests that *A. kowalevskii* had a need for periscopic vision. This highly unusual trilobite capability may indicate that it buried itself in sand or mud as a defensive behavior.

There are at least two dozen species of the *Asaphus* trilobite that have been described so far. *A. robustus*, *A. lepidurus*, *A. gracilis*, and *A. corutus* (all from Russia) each look pretty much the same as *A. kowalevskii*, including the eye-bulb design, except they have no eye stalks. The eye bulbs are attached directly to the glabella in these four species, and so if this is the mature condition for their eye scheme they would not have had the periscopic advantage.

The author speculates that these five trilobite species, and perhaps some of the other *Asaphus* species, are all one created kind and may simply represent differences due to dimorphism. The eye differences are possibly just variations within the kind. Perhaps those *Asaphus* trilobites without stalks were immature and were buried in the global Flood before they were old enough to have grown them.

Trilobite *Walliserops*

Figure 100: Replica trilobite *Walliserops* from southern Morocco [OR32].

Description: *Walliserops* trilobite replica is 3 ¼" long overall × 1 ¹⁄₁₆" wide. The trident makes up 1 ½" of the length and is ⅜" wide at the widest. The genal spines extend for ½" alongside the thorax. The thorax has 16 segments and the matrix is 4 ¼" long × 3" wide. The eyes are located high on the cephalon and are ½" center-to-center. This replica specimen does not have the spines that extend upward from the glabella and the eyes that are often reclaimed from the rock in authentic *Walliserops* fossils.

Interpretation: Relevant "18 Facts" > 1, 5, 8, 9, 10, 14. Evolutionist paleontologists have placed long-pronged-trident trilobites like this one in a different species from nearly identical trilobites with short tridents. They admit that dimorphism could also explain the differences in trident lengths. Is it possible that the male *Walliserops* trilobites jousted with their tridents? Were the tridents a sensory device or possibly a disguise? Could the trilobite lift or swing its trident? As with many other extinct lifeforms from the rock record, it is likely that scientists will never be able to definitively answer all the fascinating questions about the many marvelously designed trilobite fossils.

CHAPTER THREE
Vertebrate Animal Fossils

Archaeopteryx Displays

Figure 101: Display 'A' summarizing author's research on *Archaeopteryx*, through December, 2010.

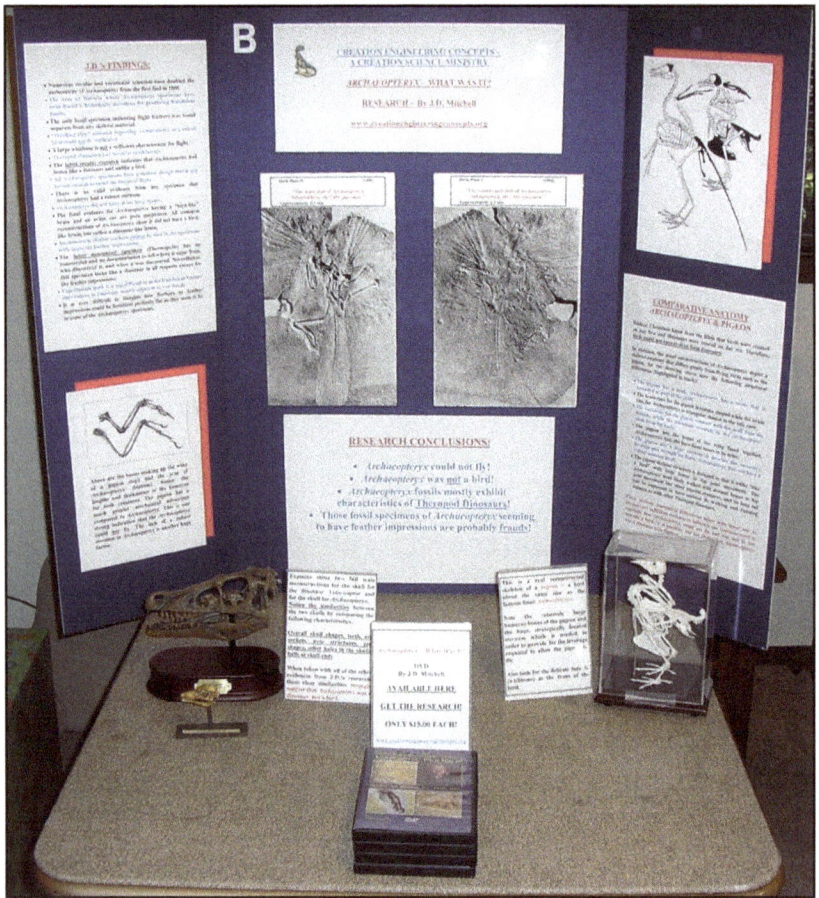

Figure 102: Display 'B' summarizing author's research on *Archaeopteryx*, through December, 2010.

Description: The author devoted a considerable amount of time for several years studying the iconic fossils of *Archaeopteryx* based upon his biblical presuppositions. The research results were summarized in the traveling two-table display shown in the two images above, and by a two-hour-long video titled, *"Archaeopteryx, What Was It?"* In 2012 an *"Archaeopteryx* Update" report was also released with new information.

The next pages of this book will show some of the related fossil replicas used in the research and will explain some of the findings of the research on this famous fossil.

Vertebrate *Archaeopteryx*

Figure 103: Full scale reconstructed skulls for *Archaeopteryx* and *Velociraptor* [DR22A, DR22V].

Description: The full-scale *Archaeopteryx* skull is 2" long × 1" wide × 1" high. The full-scale *Velociraptor* skull is 7 ½" long × 3" wide × 3 ½" high. Both skull reconstructions have very similar overall skull shapes; as well as similar teeth design, nose structures, jaw shapes, skull openings and rear-end structural balls for connecting the skulls to the cervical sockets. At first glance the *Archaeopteryx* skull looks to be a mini-*Velociraptor* skull. *Velociraptor* was a dinosaur according to all accounts.

Many evolutionists promote the idea that *Archaeopteryx* was a transitional form between dinosaurs and birds. Many creationists argue that *Archaeopteryx* was a flying bird. The author does not agree with either position.

Interpretation: Relevant "18 Facts" > 7, 8, 9, 10, 12, 13. The Bible says that birds were created on Day Five and animals on Day Six. Therefore, birds did not evolve from dinosaurs. *Archaeopteryx* and *Velociraptor* not only have similar skull structures as explained above, but are also similar with regard to many aspects of their skeletons.* The best interpretation is that *Archaeopteryx* was likely a dinosaur, not a bird, and definitely not a transitional form between dinosaurs and birds. Perhaps it was one of God's mosaics like the platypus.

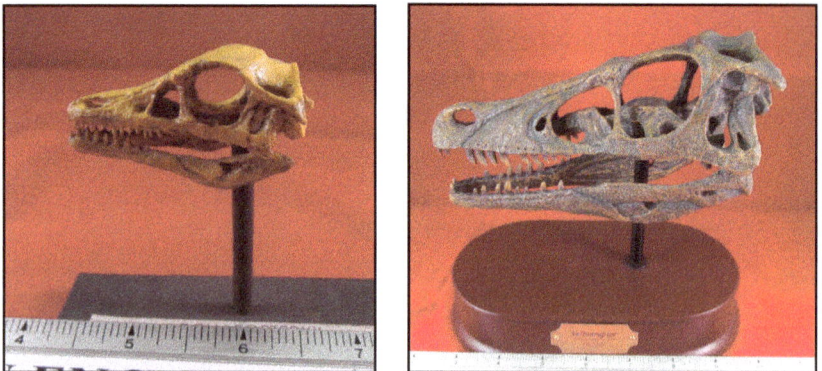

Figure 104: Side-by-side images of full-scale *Archaeopteryx* and *Velociraptor* skulls [DR22A-DR22V]

Description: The skeleton in Figure 105 was reconstructed from the parts of an actual pigeon and is about 6 ½" tall. Pigeons are about the same size as recovered *Archaeopteryx* fossils.

Interpretation: Relevant "18 Facts" > 1, 13, 14, 15. Some evolutionist and many creationist scientists promote the idea that

---

*   Wellnhofer, Peter, *Archaeopteryx-the Icon of Evolution*, Verlag Dr. Friedrich Pfeil, 2009, p. 163.

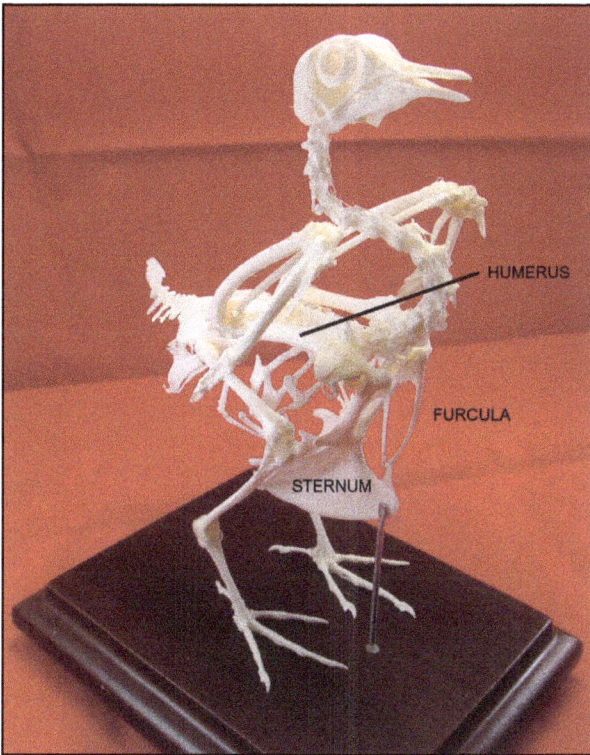

Figure 105: Actual flying bird *Columba* (pigeon) skeleton with humerus, furcula and sternum identified. This is not a fossil.

Archaeopteryx was a flying bird. This pigeon skeleton highlights important characteristics in the design of flying birds. In flying birds the humerus bones are relatively short and robust. Their furculas (wish bones) are usually thin and V-shaped and they have a massive sternum as seen above.

What is found in the *Archaeopteryx* fossils are long and thin humerus bones. The furculas are stout in cross-section and are U-shaped. *Archaeopteryx* sternums are non-existent in the fossil record. Therefore, it is the author's opinion that even if *Archaeopteryx* had feathers it could not fly and was in fact not a flying bird! Every aspect of the skeletal design of *Archaeopteryx* matches those of bipedal dinosaurs.

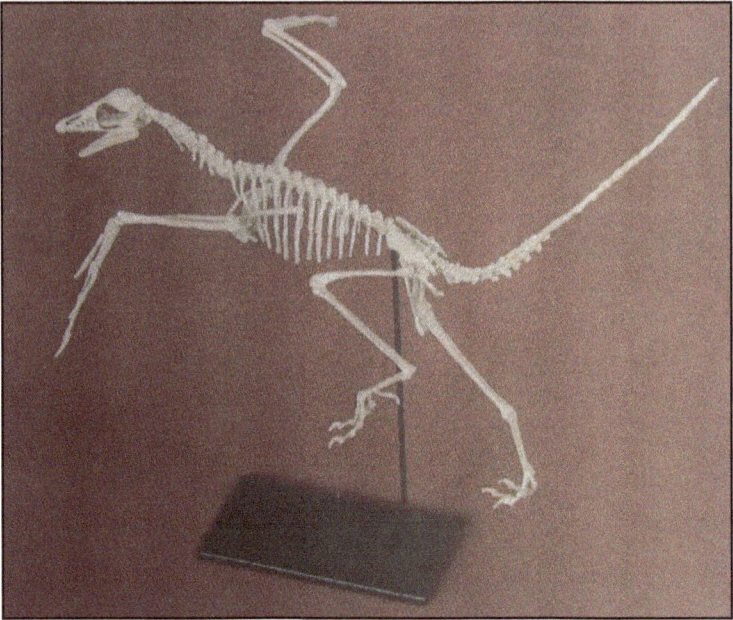

Figure 106: Reconstruction of the *Archaeopteryx* skeleton using components of multiple fossils.

Description: This reconstruction does not depict any particular specimen of *Archaeopteryx* but is a composite that shows the large differences in structural design between it and flying birds such as the pigeon.

Interpretation: Relevant "18 Facts" > 1, 13, 14, 15. Besides the points previously made regarding *Archaeopteryx* and flying bird skeletal features, notice that in this reconstruction the femurs pivot at the hips. This is another characteristic in common with bipedal dinosaurs and not birds. Birds do not have pivoting, long femurs and use the bones from the knee down to walk. Birds are "knee-walkers" not "hip-walkers."

Another characteristic of bipedal dinosaurs is the vertebrae are connected to the back of the skull just like in the *Archaeopteryx* reconstruction. In birds like the pigeon, the vertebrae connect to the bottom of the skull. A duck-bill does not make a platypus a duck and possible feather imprints in a fossil do not make *Archae-opteryx* into a flying bird!

Figure 107: Replica cast of the Berlin specimen of *Archaeopteryx* from the Solnhofen limestone of Bavaria [DR15].

Description: The Berlin specimen skull is about 2" long and the tail is about 6" long. The cast plate is 14" wide × 17 ⅜" tall. Feather imprints are present off the arms and the tail, but barbs and barbules of the feathers are not visible. There is no indication of a furcula in this particular fossil. If the skeleton had been preserved without the feather imprints, "the fossil would certainly have been identified as a small, raptorial dinosaur."*

---

* Wellnhofer, p. 75.

Interpretation: Relevant "18 Facts" > 1, 13, 14, 15. Since the first fossil of this kind was identified as dinosaur *Griposaurus* in 1862, a total of ten fossil specimens from Solnhofen have been given the genus name *Archaeopteryx*. Of those ten only three are identifiable as having accompanying feather imprints. Those three are the London specimen (originally *Griposaurus*), the Berlin specimen, and the Thermopolis specimen.

The Berlin specimen (discovered in 1876) shown in Figure 107 is the one with the least amount of disarticulation and the one shown in most of the many representations of *Archaeopteryx* as the "Icon of Evolution." The Berlin specimen is the "best" one found so far, yet when experts make comparisons of extant flying bird feathers to *Archaeopteryx* feathers they use a single fossil feather found in the limestone separate from all the skeletal specimens. This fossil feather found at Solnhofen is simply assumed to be from *Archaeopteryx*.

One possibility that has been suggested is that the feather imprints on the three aforementioned specimens are fraudulent. One strong body of evidence for this position is a record of fraudulent fossils from Solnhofen in the late 1800s. The motive for this forgery was the high monetary return for valuable fossils, especially rare fossils, at the time. This is still sometimes the case as evidenced by a number of similar frauds in the recent past of fossils from China.

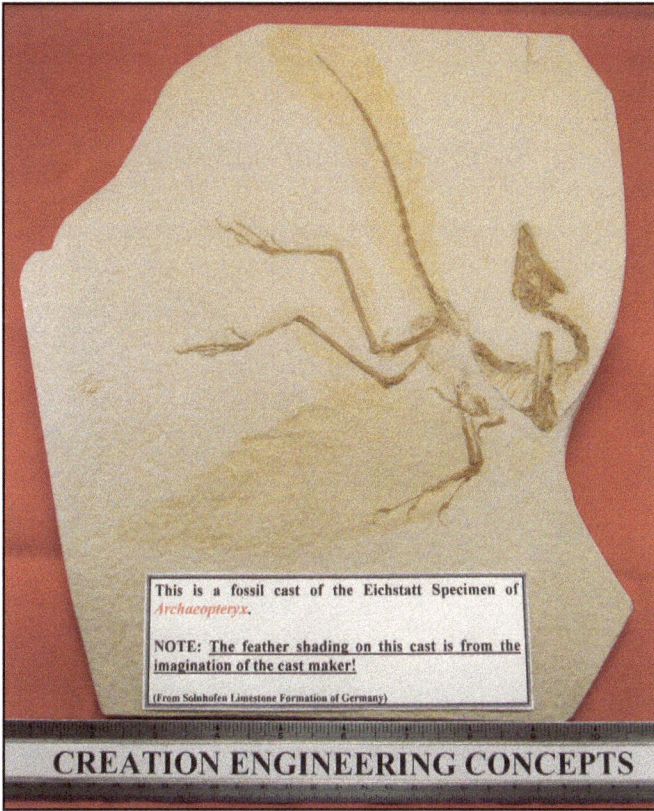

The fossil cast note reads:

This is a fossil cast of the Eichstatt Specimen of *Archaeopteryx*.

NOTE: The feather shading on this cast is from the imagination of the cast maker!

(From Solnhofen Limestone Formation of Germany)

CREATION ENGINEERING CONCEPTS

Figure 108: *Archaeopteryx* Eichstatt specimen cast from the Solnhofen limestone of Germany [DR21A].

Description: The Eichstatt skull is about 1 ⅝" long and the tail is about 5 ⅜" long. The cast plate is 10 ½" wide × 12" tall. There is no furcula present and feather imprints have been inferred by shading on the cast by the cast maker. This is the main part of the two-part fossil and the counterpart also exists and is in the author's collection, but not pictured [DR21B].

Interpretation: Relevant "18 Facts" > 1, 13, 14, 15. The Eichstatt specimen is among the top three or four so far as clearly representing the *Archaeopteryx* skeleton. Secular experts admit that the existence of feather imprints requires "interpretation" as much as examination. Again, if there are no feather imprints then this skeleton looks totally dinosaurian.

Dinosaur *Compsognathus*

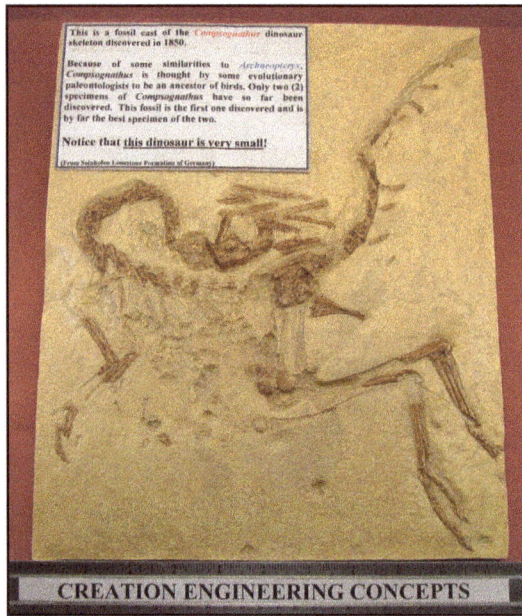

Figure 109: *Compsognathus* dinosaur replica cast from the Solnhofen strata of Germany [DR17].

Description: Cast plate is 10 ¾" wide × 12 ¾" high. *Compsognathus* is frozen in the typical "dead dinosaur" pose with its neck bent back over its body. Tail extends about 7" from the body.

Interpretation: Relevant "18 Facts" > 1, 6, 13, 15, 16. This small dinosaur fossil, which was about the same size as *Archaeopteryx*, was found in the same Solnhofen limestone strata as *Archaeopteryx*. There are a number of skeletal similarities between *Compsognathus* and *Archaeopteryx,* but the lack of any indication of feather imprints in *Compsognathus* has resulted in the two being considered entirely different life forms by most paleontologists.

Notice that this fossil is in the "opisthotonic" death pose[*] in which quite a few articulated dinosaur fossil skeletons have been found in the rock record. One explanation for this posture, also seen in several *Archaeopteryx* specimens, is that muscle spasms from failure of the central nervous system at the time of death are the cause. At any rate, it seems clear that the pose is frozen in the rock and the animal was buried rapidly by tons of water-laden sediments. This matches the biblical explanation that everything in the Genesis Flood that had the breath of life in its nostrils (i.e., all land animals) died unless they were on the ark with Noah.

London and Thermopolis specimens of *Archaeopteryx* - Furculas

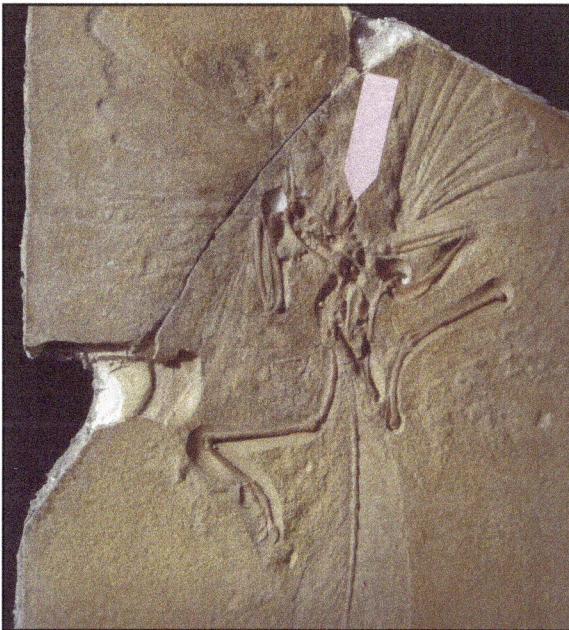

Figure 110: Photo of London specimen of *Archaeopteryx* showing furcula (arrow).

---

[*] See Mitchell, J.D., *Guidebook to North American Dinosaurs According to Created Kinds*, CEC Publications, 2014, p. 48.

Figure 111: Photo of Thermopolis specimen of *Archaeopteryx* showing furcula (notation).

Description: Both the London specimen and the Thermopolis specimen seen in the figures above have nearly identical furculas that are about ³⁄₁₆" cross-sectional diameter × 1 ⅛" long and nearly the same 'U' shape.

Interpretation: Relevant "18 Facts" > 1, 5, 14, 15. The reason for including these two photos is to show the only two *Archaeopteryx* specimens that have furculas and to show that both of these furculas are more like dinosaur furculas than flying bird furculas. From the pigeon skeleton in Figure 105 it can be seen that flying bird furculas are thin in cross-section and V-shaped. These two *Archaeopteryx* furculas are U-shaped and much more robust compared to bird furculas.

The author believes it is clear that the furculas in dinosaurs did not have the purpose of assisting in flight since there is every indication that they never did fly; and the furculas in *Archaeopteryx* look to have the same structural purpose(s) as do the dinosaur furculas.

Dinosaur *Megaraptor*

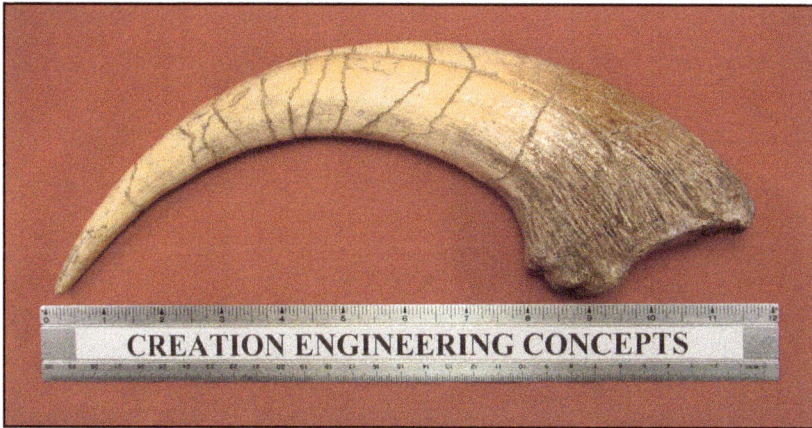

Figure 112: Replica of giant fossil claw from *Megaraptor* dinosaur from the Patagonia region of Argentina [DR14].

Description: The claw is 15" long around the curve and is 1 ⅜" thick × 3 ¾" wide at its base. For most of its length its cross-section is oval shaped.

Interpretation: Relevant "18 Facts" > 8, 9, 14. The *Megaraptor* dinosaur is known from very incomplete fossil evidence and secular experts do not agree on what kind of dinosaur it was. One thing for sure is that it had extremely large claws!

Secular scientist Alan Feduccia published the results of some measurements of various bird claws compared to *Archaeopteryx* claws and came to the conclusion from his study that *Archaeopteryx* was a perching bird and therefore arboreal. This work has encouraged a number of influential creationists to write that *Archaeopteryx* was therefore a bird. However, the author was not able to replicate Feduccia's work and points out that birds perch on their feet, not on their claws. Even world-renowned *Archaeopteryx* expert Peter Wellnhofer wrote, "In conclusion, there is no evidence for the assumption that *Archaeopteryx* was arboreal..."*

---

* Wellnhofer, p. 144.

Finally, the author points out the great similarity between the claws of *Archaeopteryx* and this *Megaraptor* claw, *Velociraptor* claws, and *Utahraptor* dinosaur claws. There is large variation in size between these various claws, but the design and shape are very similar for all. This fits with the hypothesis that *Archaeopteryx* was not a bird, but was either a dinosaur or a mosaic. The reader can see the similarity in these claws by examining the fossil images shown in Figures 106, 107 and 108.

Dinosaur Furculas

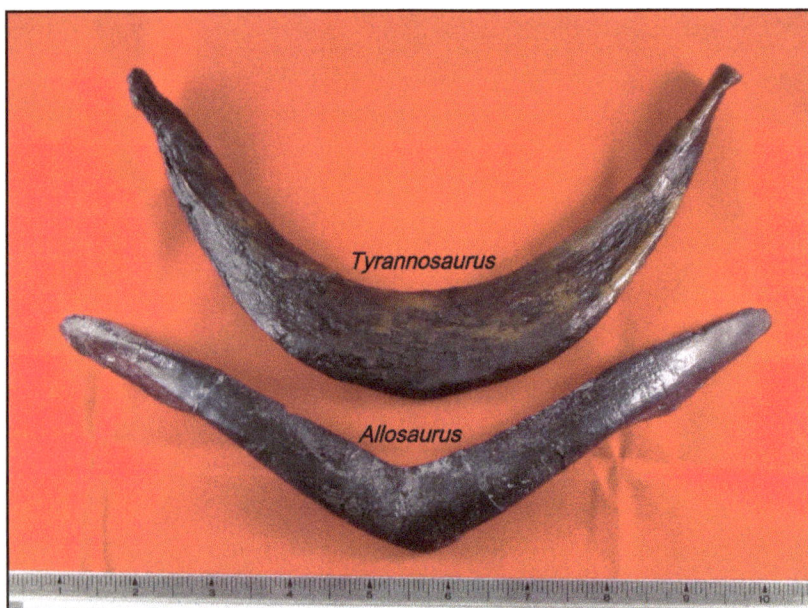

Figure 113: *Allosaurus* and *Tyrannosaurus* furculas [DR20 & DR24].

Description: The *Allosaurus* furcula DR20 is a flat 'V' shape with an overall length of eleven inches around the inside of the curve. At maximum cross-section it measures 1" × 1 ½". At each end of the furcula is a facet for contact with the shoulder blades.

The *Tyrannosaurus* furcula DR24 is boomerang shaped with an overall length around the inside curve of 11 ½." At maximum

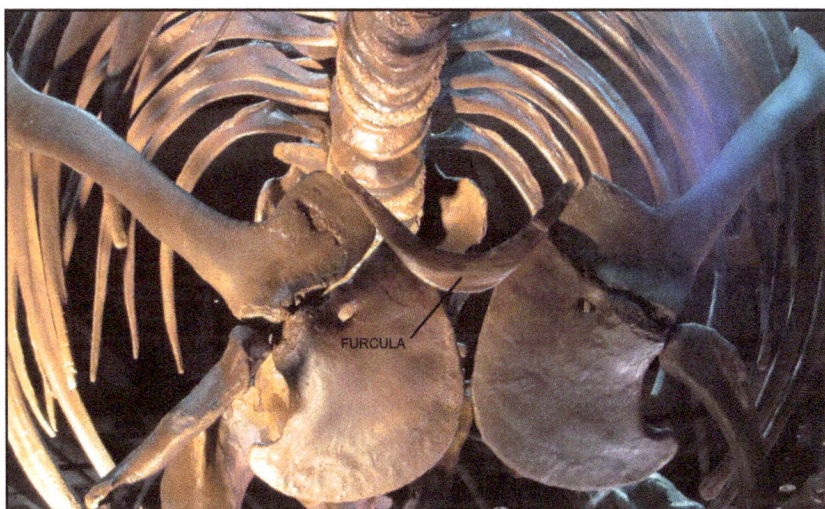

Figure 114: Typical reconstruction of a Tyrant Bipedal Dinosaur showing chest and furcula [Houston Museum].

Figure 115: Photo of *Allosaurus* reconstructed skeleton with furcula noted [Cleveland-Lloyd Quarry].

cross-section it measures 1 ½" × 1 ½." The furcula has facets on each end that are less defined than in the *Allosaurus* furcula. The author points out that the boomerang-shaped furcula of

Archaeopteryx corresponds closely to that of the Tyrant Bipedal dinosaurs, but not to that of flying birds.

Interpretation: Until fairly recently, furculas were thought to be found only in birds. Therefore, evolutionists assumed the London specimen of *Archaeopteryx* with a furcula was an indicator that it was a bird or on the way to evolving into a bird. It is now known that every skeletal characteristic of *Archaeopteryx* closely matches the design of Tyrant Bipedal (theropod) dinosaurs.* The only remaining characteristic that is bird-like are the feather imprints found in a few of the Solnhofen limestone fossils.

So, that leads the author to the conclusion that *Archaeopteryx* was not a transitional form, nor a flying bird, but most likely a dinosaur with feathers fraudulently added or a mosaic like the platypus.

Dinosaur Brain Endocasts

Descriptions: The *Allosaurus* endocast (Figure 116) is 8 ¼" long × 4 ¾" high × 2 ⅜" thick. The *Tyrannosaurus* endocast (Figure 117) is 7 ¼" long × 2 ¾" wide × 2 ¼" thick. Both endocasts are thought to be from adult animals. Notice that the *Allosaurus* endocast is larger than the *Tyrannosaurus* even though larger fossil skulls have been recovered from *Tyrannosaurus* remains. Both dinosaurs are categorized as Tyrant Bipedal created kind by the author.**

Interpretation: Relevant "18 Facts" > 8, 9, 10, 12, 14. An endocast is the result of infilling the brain cavity of an animal's skull. This infilling can be natural where the cavity is filled with fine grained sediments associated with the fossilization process. Or, scientific endocasts can be made from the artificial infilling by researchers to obtain a cast that uses the skull itself as a mold.

One assumption here is that the brain cavities reflect the actual

---

* Wellnhofer, p. 163.

** See Mitchell, J.D., *Guidebook to North American Dinosaurs According to Created Kinds*, CEC Publications, 2014.

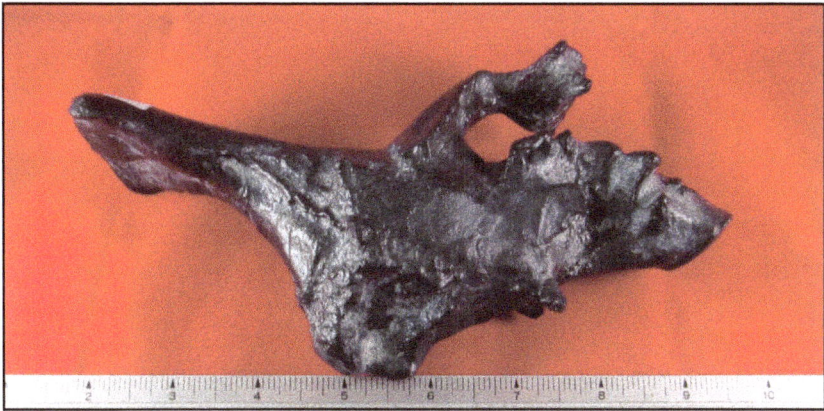

Figure 116: *Allosaurus fragilis* brain endocast [DF11].

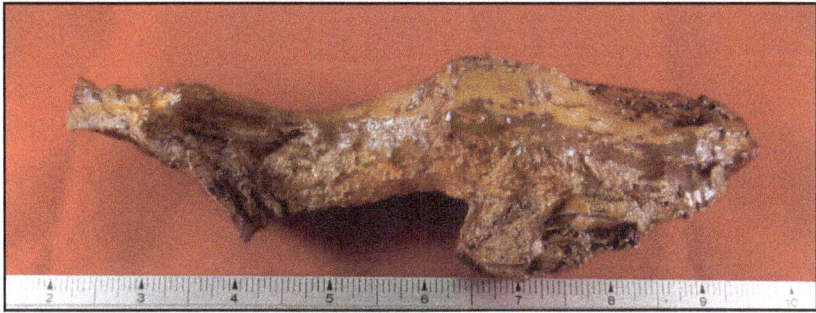

Figure 117: *Tyrannosaurus rex* brain endocast [DF12].

sizes of the brains. In fish the brain may fill as little as fifty per cent of the brain cavity. In crocodiles and mammals the brains almost completely fill the skull cavities. Since dinosaurs are similar to crocodiles and mammals, let us assume that endocasts DF11 and DF12 do realistically represent casts of the brains for these two individuals. Does this mean that *Allosaurus* was smarter than *T-rex*?

This is unlikely because it is very difficult to make these endocasts since the skulls are usually badly damaged or crushed, and if they have natural endocasts they are nearly impossible to remove from the skulls. These two endocasts appear to be examples of variation within the Tyrant Bipedal kind. Using biblical presuppositions we can believe that God designed these brains to be the perfect size, design, and construction to allow the animals to operate and adapt as needed while living on earth.

Tyrant Bipedal Dinosaur Teeth

Figure 118: Fossil replica teeth from Tyrant Bipedal dinosaurs. *Tyrannosaurus* with root is in front [DR19], *Tyrannosaurus* with jaw fragment is at left rear [DR1], and *Allosaurus* with root is at right rear [DR32].

Description: DR19 is 11" long around the curve × 2 ⅛" maximum width × 1 ⅜" maximum thickness. DR19 tooth less the root is 3 ¼" long. DR1 is 6" long × 1 ¾" maximum diameter at its base. DR32 is 5 ¼" long around the curve × ¹⁵⁄₁₆" maximum width × ½" maximum thickness. DR32 tooth less the root is 2 ⅛" long.

Interpretation: Relevant "18 Facts" > 8, 9, 10, 14. The author has inspected over fifty fossil and replica Tyrant Bipedal dinosaur teeth and has concluded that the teeth are quite similar from genus to genus. Notice in the figure how similar the *Tyrannosaurus* and *Allosaurus* rooted tooth designs appear. Both teeth have similar geometry and comparable grooves caused by the crowns of erupting teeth below them. The main difference is in the size, which can be explained by the assumed age difference between the two specimens where one is an adult and the other a juvenile.

According to evolutionary theory, these two genera were on earth millions of years apart and so are only related in some deep-time evolved manner. The biblical creationist has an entirely different interpretation. Since both *Tyrannosaurus* and *Allosaurus* fossils are in the rock record as a result of the worldwide Flood, they lived at the same time. They were created along with the other land animals and man on Day Six of God's creation as explained in the Bible. At the time of the fall, God instituted the curse, and for the 1,600 year period from the fall to the Flood animals, including dinosaurs, developed variations within the kinds (as allowed by their DNA) that are seen today in the fossils. These variations are not macroevolutionary changes from one created kind to another kind. Therefore, *Tyrannosaurus* and *Allosaurus* are both of the same Tyrant Bipedal created kind.

Figure 119: Tyrant Bipedal dinosaur artistic reconstruction by Marianne Pike.

Dinosaur Foot Claws

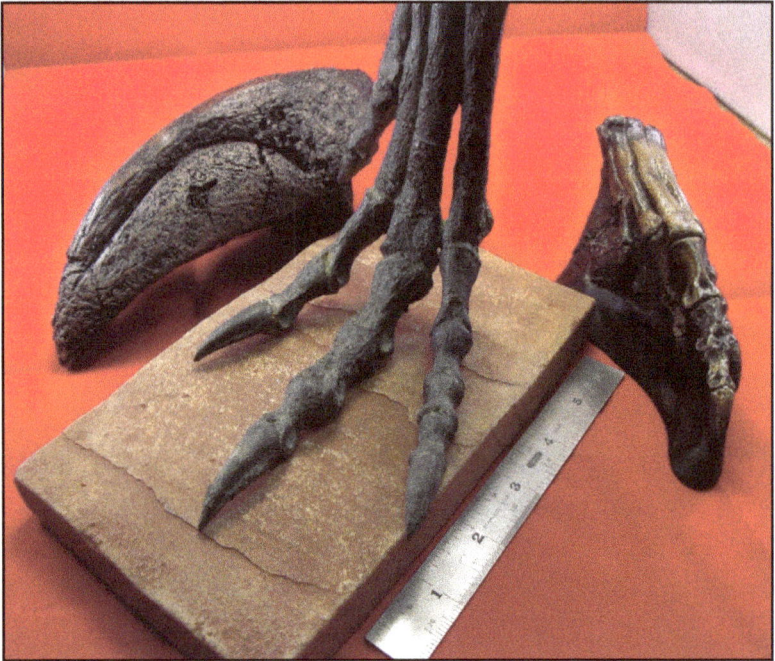

Figure 120: Replica foot claws for Bipedal Browser and Tyrant Bipedal dinosaurs. *Tyrannosaurus* is at left [DR4], *Othnielia* is in center [DR25] and *Nanosaurus* is at right [DR9].

Description: DR4 claw is 9" long around the curve × 3 ⅜" high at the base × 3 ⅜" wide at the base. The claw on the longest toe of DR25 is 1 ½" long around the curve × ⅜" high at the base × ½" wide at the base. The claw on the longest toe of DR9 is 1" long around the curve × ¼" high at the base × ⁵⁄₁₆" wide at the base. (*Othnielia* and *Nanosaurus* are two genus names for the same Bipedal Browser dinosaur.)

Interpretation: Relevant "18 Facts' > 8, 9, 10, 13, 14. The Bipedal Browser dinosaurs such as *Othnielia* had teeth that indicate they were plant eaters at the time of the Flood. The Tyrant Bipedal dinosaurs such as *Tyrannosaurus* left evidence that indicate they were carnivorous at the time of the Flood. Yet the claws on their feet were very similar in design. For what purposes did dinosaurs use their claws?

At the time of the creation the Bible tells us that all animals and mankind ate only plants (Genesis 1:29). Therefore, we can know that an original purpose for dinosaur claws was not to tear the flesh of other animals. That use came only after the fall and the institution of the curse by God. In the beginning the claws of all animals were designed primarily to protect the toes and help with the harvesting and ingestion of plant matter. The claws would have also been helpful for nest building.

Dinosaur Femur Bones

Figure 121: Tyrant Bipedal dinosaur (*Allosaurus*) juvenile replica fossil femur bone from the Cleveland-Lloyd Quarry in Emery County, Utah [DR34].

Description: The replica *Allosaurus* femur bone is 9 ½" long. At the center of the shaft the bone is about ⅞" diameter. The femoral head where the bone fits into the hip is also about ⅞" diameter. The maximum bone width at the knee joint (left end) is 1 ⅞."

Interpretation: Relevant "18 Facts" > 1, 2, 3, 8, 9, 10, 14, 15. Since the femoral head is 90 degrees to the centerline of the bone shaft

Figure 122: Display of *Allosaurus* femur bones from the Cleveland-Lloyd dinosaur bone quarry in Utah.

we can deduce that the *Allosaurus* walked with its legs directly under its body.

At the Cleveland-Lloyd quarry are more than 12,000 individual dinosaur bones that have been discovered. The bone bed map for the quarry shows that these bones are completely disarticulated and represent a number of different dinosaur kinds. This situation corresponds to a biblical creationist interpretation that the quarry is a true fossil bone jumble and exists as a result of the Genesis Flood cataclysm.

Many bones from *Allosaurus* have been recovered from the quarry including a large number of femurs. Some of the femurs are shown in the figure above. Notice that there is a variation in size as well as in the detailed formation of the bones. It seems reasonable to assume these femurs represent gender and age dimorphism, as well as general variation within a kind, over the 1,600 years from the time of the fall to the onset of the Flood. The replica juvenile femur DR34 is about ⅕ the length of the largest *Allosaurus* femurs recovered so far.

It is the practice of those recovering the bones at Cleveland-Lloyd to depict all of the bone found there as being black. So they paint them all black when displayed. In reality, the bones are in the rock close to the same color as the matrix which is a brownish-gray color. By painting all the bones black, cast replica bones can be used along with real bones in skeletal reconstructions. Not all museums paint their dinosaur skeletal reconstruction components black, but most do choose a particular color common in their displays, with shades of brown and gray being other popular colors.

Allosaurus Radius Bones

Figure 123: Pathological (above) and normal replica fossil radius bones from *Allosaurus* dinosaurs [DR35].

Description: The pathological radius is 6 ¾" long and the thickened callus on it is about 1 ½" × 1 ½". The normal radius bone is 7 ⅛" long × about ¾" diameter at the midpoint.

Interpretation: Relevant "18 Facts" > 1, 2, 3, 12, 14, 16. Paleopathology is the study of diseases and injuries in fossils. More than a few dinosaur fossils have been discovered with these defects. Any fossil that shows signs of pre-death disease or injury is called "pathological."

The shape of the two radius bones above tell us that they are from the left arms of two different animals. The left ends of the bones are shaped to connect to the humerus bones. All around the pathological bone, rough and porous bone formed a callus after it was broken and as it healed. We can expect that this was very painful for the *Allosaurus* and it would not have been able to use its left arm until it healed. Since the healed bone is crooked the dinosaur would have been "crippled" and the callus may have rubbed against the other arm bone, the ulna.

The Bible tells us that there was no death, disease or suffering before the fall. Death came from man's sin, so that means that dinosaurs found in the rock record (even those with pathologies) were buried in the Flood and became fossilized after the fall. The *Allosaurus* broke its arm sometime during the 1,600 year period between the fall and the Flood. (Of course, some dinosaurs were taken onto Noah's ark and were not buried in the Flood. Their bones and the bones of their descendants were seldom if ever fossilized, according to the biblical creationist hypothesis.)

Edmontosaurus Toe Bones

Description: The larger DF9 bone (Figure 124) would be located adjacent to the metatarsal bone in the foot of the dinosaur and is 2" long × 2 ½" wide. The next adjacent phalange DF10 is 1" long × 1 ⅞" wide. The size and color of the fossils indicate that they may be from the same foot on the same dinosaur.

Interpretation: Relevant "18 Facts" > 1, 2, 3. Figures 124 and 125 illustrate the difficulty involved in accurately reconstructing

Figure 124: *Edmontosaurus* fossil phalanges (toe) bones from the Lance Creek formation of Wyoming [DF9 & DF10].

extinct vertebrate animals from their fossilized remains. Usually the bones are found in bone jumbles and numerous components of a complete animal are found to be missing. For example, the author has seen a number of reconstructions of the *Edmontosaurus* foot and each one depicted the amount of toe splay and the distance between phalanges differently. Without having a live animal to look at it is impossible to determine these details with certainty from fossil bones alone.

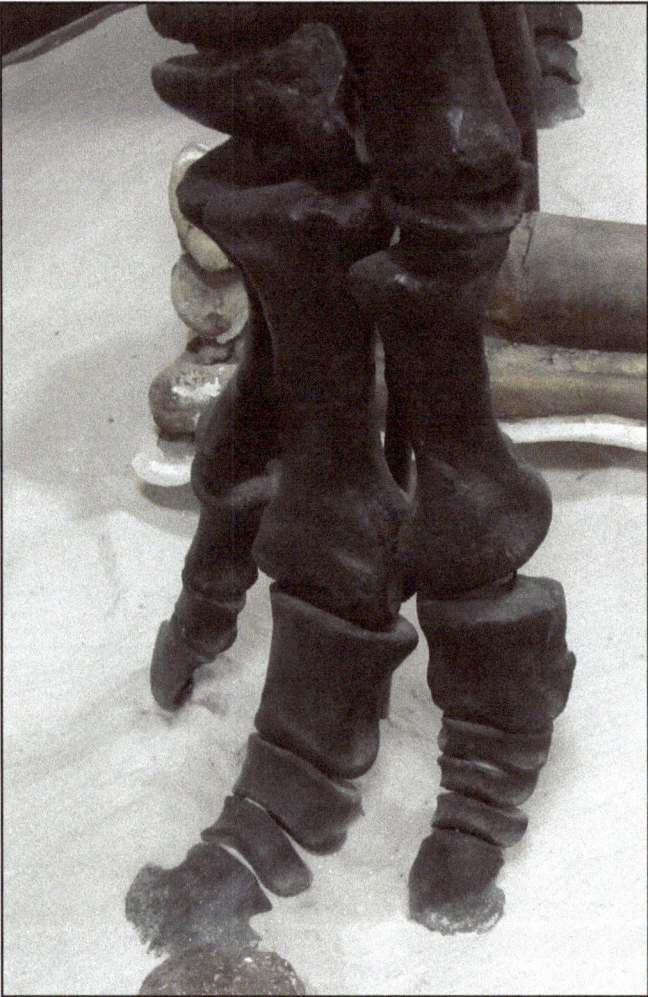

Figure 125: *Edmontosaurus* dinosaur foot reconstruction at the South Dakota Museum of Geology.

Duck-Billed Dinosaur Vertebrae

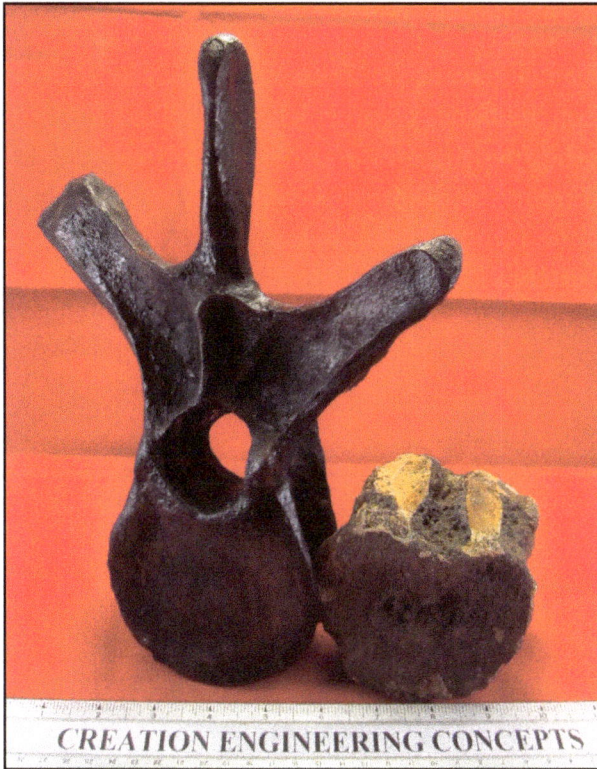

Figure 126: Replica complete lumbar vertebra (left) from the Hell Creek formation of southern North Dakota and partial fossil lumbar vertebra from the Hell Creek formation of Montana. Both vertebrae are from Duck-Billed dinosaurs [DR28 & DF1].

Descriptions: The replica vertebra DR28 is from an *Edmontosaurus* dinosaur and is 11" tall × 8" wide × 3" thick. It has the vertical neural spine, two nearly complete transverse processes, and a clear neural canal which is the hole for the spinal cord. The vertebra was recovered along with three other lumbar vertebrae and several other bones of an incomplete disarticulated skeleton estimated to represent about sixty per cent of the animal.

The fossilized vertebra DF1 has lost its neural spine and transverse processes, but it is obvious where they originally were solidly attached to the centrum. The fossil vertebra is 4" in diameter × 3" thick and is similar enough in design to be from the same genus of dinosaur as the more complete replica.

Interpretation: Relevant "18 Facts" > 1, 2, 3, 8, 9, 10, 14, 16. The excavators of the bones from which the replica vertebra was made explained the fossilization and recovery of the partial *Edmontosaurus* as a many "millions of years" process. Each of the bones was in a bone jumble and was broken or distorted. The partial fossilized vertebra DF1 is badly damaged. Local flooding and other uniformitarian geological actions can be used to explain these fossils over deep time, but the worldwide Flood just 4,500 years ago is a more satisfying explanation to an open and logical mind.

Figure 127: Duck-Billed dinosaur artistic reconstruction by Marianne Pike.

Duck-Billed Dinosaur Teeth

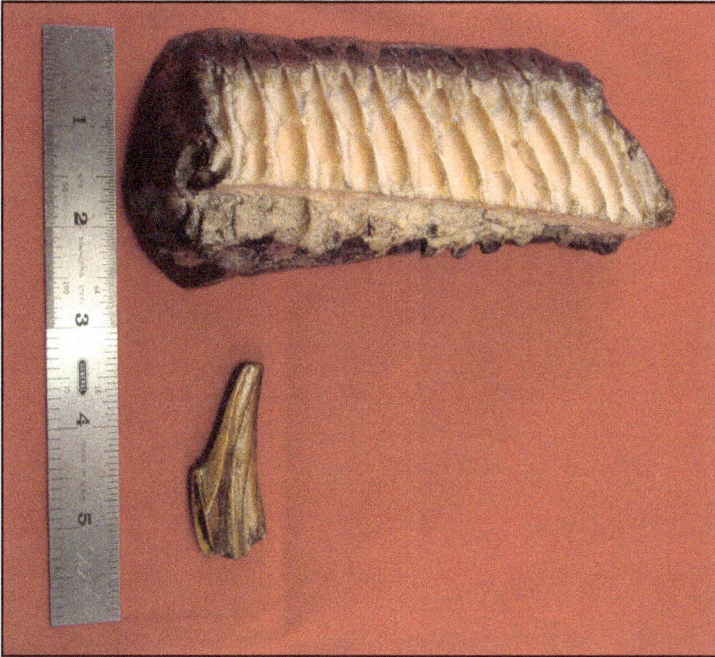

Figure 128: Duck-Billed dinosaur (*Edmontosaurus*) replica tooth row and replica tooth from western North America [DR30 & DR29].

Description: DR30 Tooth row is 2 ½" wide × 2" tall × 5 ½" long. Individual teeth in the row are 2 ⅝" long × ⁵⁄₁₆" wide. The DR29 loose tooth replica is 2" long × ¼" wide. The tooth row looks to be from a larger dinosaur than does the loose tooth.

Interpretation: Relevant "18 Facts" > 8, 9, 10, 13, 14. In the Duck-Billed dinosaurs the teeth were made up of closely packed rows of self-sharpening teeth. That design allowed for the teeth surfaces to be renewed as they were worn. These teeth would be good for shearing and grinding tough plant material.

According to evolutionary theory, all dinosaurs have a common ancestor and the various tooth designs evolved over millions of years from that unknown ancestor. A much more satisfactory hypothesis

Figure 129: Photo of an actual fossil *Edmontosaurus* jaw showing its tooth battery [Museum of the Rockies]

is that the different tooth designs were created in the beginning according to the foreknowledge of the Ultimate Designer.

Dinosaur Skin Imprints

Figure 130: Fossil replica skin imprints for a Duck-Billed dinosaur (left) and a Horn-Faced dinosaur [DR5 & DR12].

Description: The Duck-Billed dinosaur imprint DR5 is 3 ⅝" wide × 5 ¾" long. The largest scale is ¼" × ⅜" and the smallest scales are about ¹⁄₁₆" square.

The Horn-Faced imprint DR12 is 3 ¼" wide × 4 ¾" long. The largest scale is 2 ¼" in diameter and the smallest scale is about ⁵⁄₁₆" square. The colors of the replicas were chosen by the cast makers and no scientist really knows the color(s) of the skin of any dinosaur.

Interpretation: Because many evolutionists believe birds evolved from dinosaurs, it is important for them to assume that many dinosaurs had feathers. Therefore, the existence of feathers should be obvious in the dinosaur fossils.

As is clear in these two replicas, the skin of dinosaurs had scales similar to those found on today's reptiles. The structure of feathers is extremely more complex than the structure of scales, so feathers could not have evolved from scales no matter how much time is allowed. While God could have, and may have, designed some dinosaurs with feathers, there is little real evidence for this in the few fossil dinosaur skin imprints that have been so far discovered.

Triceratops Horns

Description: Each horn in Figure 131 is 22" long and the distance from tip to tip is 20." At their base each horn is 4 ⅝" in diameter and the distance between horns where they connect to the skull is 4 ½." The skull segment section is 6 ½" wide × 14" long.

Interpretation: Relevant "18 Facts" > 8, 9, 10, 13. Just prior to the Flood, we can interpret the various frill designs and horn arrangements in Horn-Faced dinosaurs to possibly have been adapted for defensive purposes. Soon after creation, before the fall, perhaps the horns were used for various food gathering purposes since there would have been no need for defense in the perfect creation prior to man's sin. Another possible use during both periods could have been for attracting mates.

There are about twenty genus names currently assigned by

Figure 131: Replica fossil double brow horns of the Horn-Faced dinosaur *Triceratops* from western North America [DR2].

secularists to the Horn-Faced dinosaurs because of the variations in head adornments that have been discovered. The skeletal design in all these genera is otherwise pretty much the same, consequently the differences in horns and frills is likely due to variations within a single created kind. These variations came to be due to age and gender differences and genetic adaptations that happened after the institution of the curse and prior to the Flood.

Triceratops Teeth & Bones

Description: The *Triceratops* tooth segment in the gem jar of Figure 132 is ⅞" long and the adjacent frill segment is 1 ½" × 1 ½" × ⅝" thick. The other fossil bones were recovered in the same general area where other *Triceratops* fossils were found, but could be from other animals.

Figure 132: Fossil tooth, frill section and other bone parts of *Triceratops* from the Hell Creek formation of Montana and South Dakota [DF8].

Interpretation: Relevant "18 Facts" > 1, 2, 3. The condition of these fossils can be understood if it resulted from a cataclysmic worldwide Flood. These bones were all rapidly buried and parts were torn asunder by the unimaginable forces of the Flood over a period of one year. Some of the damage to the fossils undoubtedly occurred due to erosional forces as they were uncovered from the

Figure 133: Horn-Faced dinosaur artistic reconstruction by Marianne Pike.

Figure 135: Photo of a complete *Triceratops* skeletal reconstruction from above showing the ribs connected to the vertebrae [Houston Museum of Natural science].

rock. It is highly likely that much fossil material ended up far from where it was when the deluge first affected the associated life forms at the time of Flood onset.

Triceratops Rib

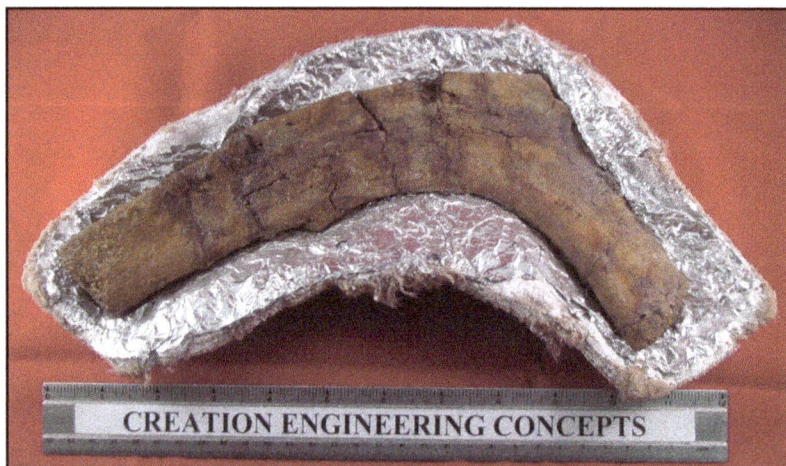

Figure 134: Rib section from *Triceratops* dinosaur displayed in the plaster jacket used to protect the fossil during its recovery from the Hell Creek formation of Montana [DF7].

Description: The rib section is 15" long around the outside edge × 2 ⅞" wide × 1" thick. There are several cracks that have been filled with glue to hold the fossil together.

Interpretation: Relevant "18 Facts" > 1, 7, 8, 10. This fossil rib section looks to be from an adult specimen. The remaining section of the rib was not found. The section is 25 to 35 percent of the complete rib which was on the order of four feet long around the curve. The skeletal features of *Triceratops*, *Styracosaurus*, *Chasmosaurus*, *Torosaurus* and other Horn-Faced dinosaurs were all very similar with genera differences mostly in the skulls. This lends credence to the biblical creationist idea that all were just one created kind.

Triceratops Fibula

Description: The bone is 3" wide × 1" deep × 2 ¼" long encased in a 5" × 6" × 2" fine sandstone matrix. The exterior of the bone is permineralized to a thickness of ⅛" to ⅜" thick and the rest of the bone is missing. This was identified as a partial *Triceratops* fibula bone by paleontologists at the Black Hills Institute in Hill City, South Dakota.

Figure 136: Partial fossilized fibula (calf) bone of *Triceratops* from the Hell Creek formation of South Dakota [DF4].

Interpretation: Relevant "18 Facts" > 1, 15. Fossil bones are found in a number of different conditions of preservation from nearly 100 percent original bone to 100 percent totally permineralized. They are found in mudstone, sandstone, limestone and every combination of minerals that make up the earth's sedimentary layers. Yet, where do we see animal bones being fossilized today? Not in lake beds, rivers, or oceans as hypothesized by secularists. It makes more sense that most fossil bones, especially dinosaur bones, were the result of the special conditions of the one-time global event at the time of Noah as described in Genesis.

Dinosaur Leg Bone Section

Description: The bone section is totally permineralized so that it is extremely hard and very heavy, weighing three pounds. The bone marrow has been largely replaced by the red mineral jasper. The specimen cross-section is 'D' shaped and is 4 ⅛" wide × 3" high. By

Figure 137: Fossil dinosaur leg bone section from Utah [DF5].

extrapolation it can be estimated that the original cross-section of the bone was about 5" wide × 3 ¼" high. The bone section is 3" long.

Interpretation: Relevant "18 Facts" > 1, 3, 15. The characteristic D-shaped cross section and the size of the bone indicate it is likely from a fairly large dinosaur leg bone. Hardened fossil dinosaur bones are often found broken into pieces like the specimen as a result of the hydraulic, volcanic, and tectonic actions of the Genesis Flood; and the subsequent geologic forces and weathering applied to them as they erode out of the layers they were initially formed in. The photo below is of a cabin that was made of nearly 6,000 individual bones similar to the leg bone above.

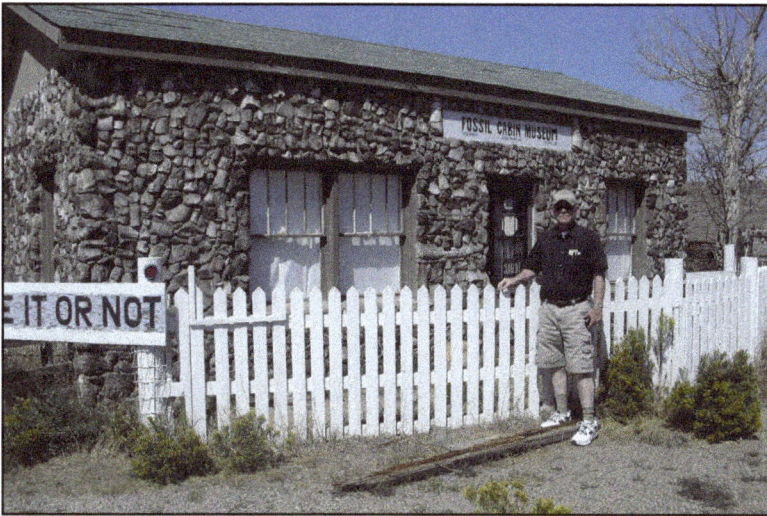

Figure 138: Fossil cabin near Como Bluff in Wyoming with walls made of 5,796 individual rock-hard dinosaur bone pieces.

Dinosaur Bone Marrow

Description: The cone-shaped specimen DF3 in Figure 139 is 1 ½" diameter at the base × 5 ½" long. DF5 was previously described. These two pieces are very dense. The other pieces are of varying densities, but less than DF3 and DF5.

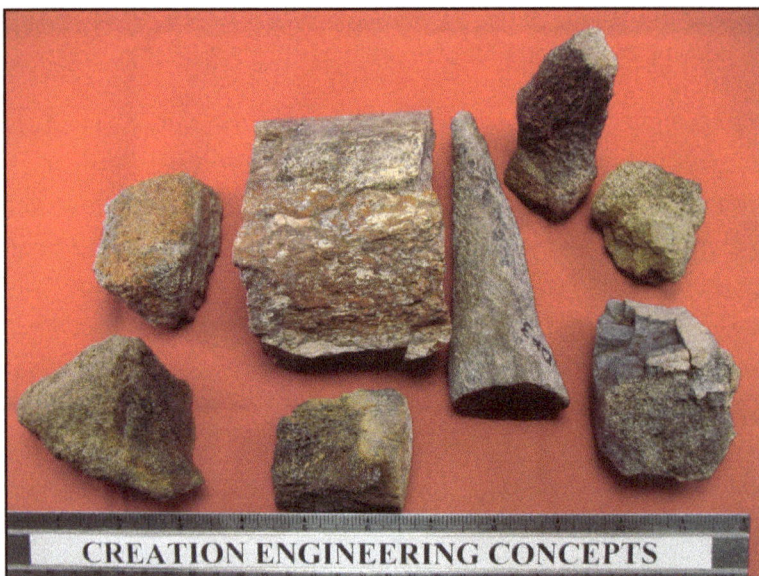

Figure 139: Eight examples of fossilized dinosaur bone marrow. Cone-shaped specimen [DF3] is from the North Saskatchewan River, Edmonton, Alberta. The largest piece is specimen DF5 from Utah. The other six are from the Hell Creek formation of Montana.

Interpretations: Relevant "18 Facts" > 1, 2, 3, 4, 14. The bones of living animals are one-third water, have blood vessels going in and out, have nerves that can feel pressure and pain; and certain bones contain marrow which produces blood cells. The bone marrow is contained inside a wide layer of strong compact bone and is a spongy material.

When one is looking for fossils in the field, one way to differentiate bone from rock is to look for the characteristic bone marrow morphology. The density and color of fossilized bone marrow is determined by the type and extent of the minerals that have replaced the original tissues of the bone. However, in most cases the characteristic "spongy" look of the fossil is what identifies the marrow.

Bones found lying around in nature lose the interior marrow component long before the harder exterior portion deteriorates. Therefore, fossilized bone marrow is always an indication of complete and rapid burial. The biblical creationist understands that the uniformitarian scenario for fossilization, one of slow burial over

Figure 141: Plate-Backed dinosaur artistic reconstruction by Marianne Pike.

long periods of time, does not match reality in any way. Vertebrate bones of today and yesterday have a common design for marrow because of the work of the Common Designer.

Stegosaurus Tail Spike

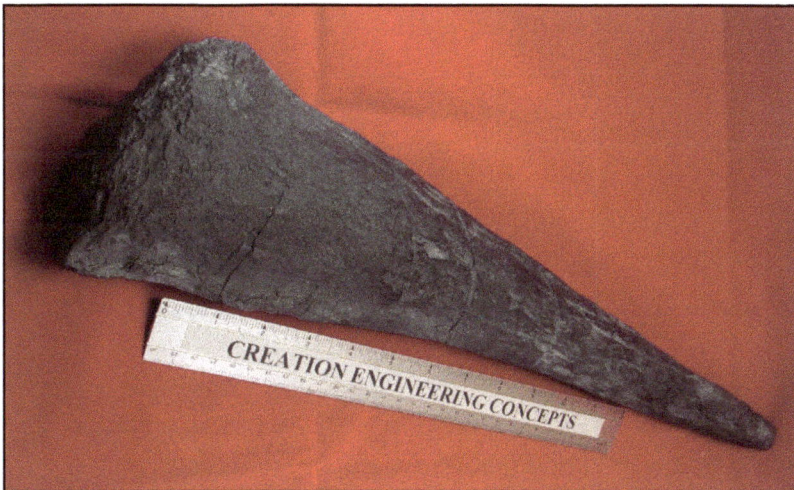

Figure 140: Fossil replica Plate-Backed dinosaur (*Stegosaurus*) tail spike from the Cleveland-Lloyd bone quarry in Utah {DR31}.

Description: The tail spike is 20" long × 2 ½" thick × 5" wide at its base. It is ¾" wide × ½" thick at the tip. The numerous grooves in the surface of the spike were probably for blood vessels.

Interpretation: Relevant "18 Facts" > 1, 2, 3, 15. Plate-Backed dinosaur reconstructions are usually depicted with four of these tail spikes at or near the end of the tail. The spike from which this replica was made is from the Cleveland-Lloyd quarry in Utah where hundreds of dinosaur bones have been uncovered in a huge bone jumble.

   The biblical creationist agrees that these spikes were undoubtedly used for defense by these dinosaurs after the fall. There is no agreement for how the jumble came to be however. The evolutionist must assume some sort of local flood action while the biblical creationist would say the bone jumble is a result of the worldwide Flood.

Stegosaurus Bony Tail Plate

Figure 142: Plate-Backed dinosaur (*Stegosaurus*) tail plate replica fossil [DR33].

Description: This bony plate from the tail of a Stegosaurus is 2 ¾" high × 7 ¼" long × 1 ⁹⁄₁₆" thick at the base. Its cross section is triangular in shape. Largest groove in the surface is ¹⁄₁₆" wide × ¹⁄₁₆" deep × 3" long.

Interpretation: Relevant "18 Facts" > 1, 2, 3, 14. This is one of the smallest tail plates on what was likely an adult *Stegosaurus*. The plates on the center back were much larger than this one. The groove on the side indicated a horny sheath that may have covered the plate and holes at the base allowed for blood to be pumped into the plate. The plates were anchored to the back of the animal in two parallel rows.

These bony plates probably were for heat regulation and/or to attract mates. After the fall, the plates undoubtedly served defensive purposes as well. Among the dinosaurs the Plate-Backed kind exhibited some of God's most unique designs.

Camarasaurus Tooth

Figure 143: Replica fossil tooth from a Long-Necked Boxy-Headed kind of dinosaur (*Camarasaurus*) from the Morrison formation of Utah [DR26].

Description: The tooth is 4 ⅛" long overall × 1 ¼" wide × ¹⁵⁄₁₆" thick. Working end of the tooth is slightly cupped and is 1 ¼" wide × ⅝" thick. The worn area on the tip of the tooth is ¾" long × ⁵⁄₁₆" wide. The tooth root is 2" long × ¾" wide × ⅞" thick.

Interpretation: Relevant "18 Facts" > 8, 9, 10, 14. An analysis of the teeth and the skull of this kind of dinosaur has resulted in an understanding that the teeth were designed to clip off plant

Figure 144: Long-Necked Boxy-Headed dinosaur artistic reconstruction by Marianne Pike.

Figure 145: Long-Necked Boxy-Headed dinosaur (*Camarasaurus*) skull with good view of its teeth [Black Hills Institute].

matter using a chomping action. The vegetable matter would then go directly into the animal's digestive system since there were no grinding teeth available for chewing. These dinosaurs probably had gizzards with stones inside to help with digestion.

On the replica tooth is a pattern of wear where its tip contacted an opposing tooth. The flat worn area can be seen in the figure between the tip of the tooth and the 1" mark on the scale. The biblical creationist author has hypothesized from the tooth evidence that the eating system for this kind of dinosaur did not change very much over the 1,600 years from the time of creation to the Flood.

Brachiosaurus Toe Bone

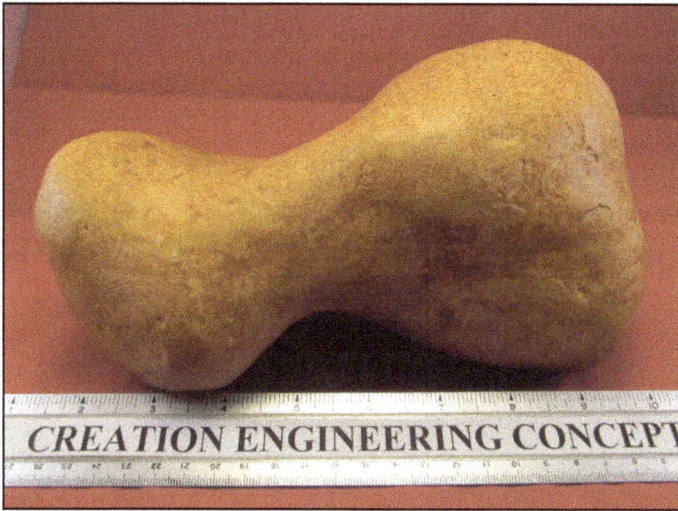

Figure 146: Replica of a Long-Necked Boxy-Headed dinosaur (*Brachiosaurus*) toe bone [DR13].

Description: This massive replica toe bone is 9" long × 4 ½" wide × 4 ⅝" maximum height.

Interpretation: Relevant "18 Facts" > 8, 9, 10, 13, 14. Most reconstructions show the toe bones of these large dinosaurs in a near vertical position, as seen in Figure 147. This foot design is similar

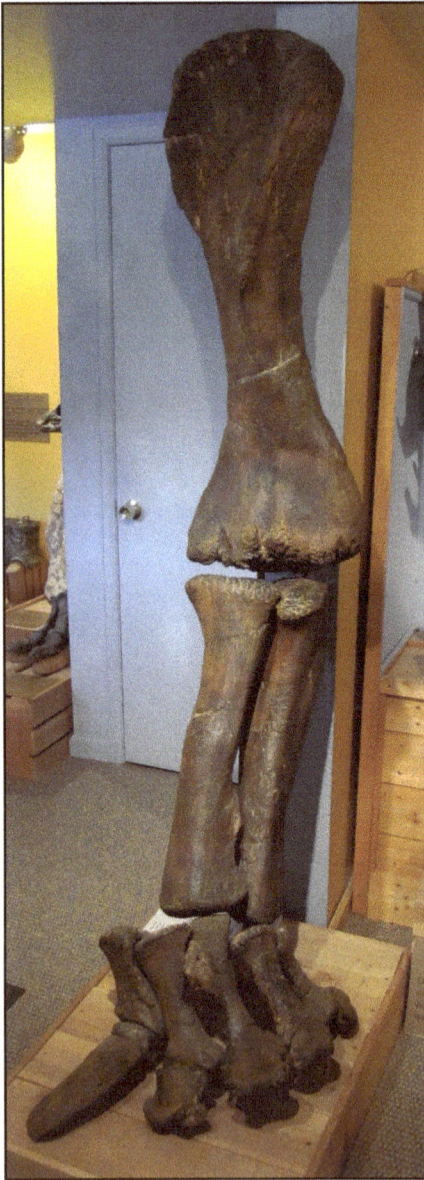

Figure 147: Reconstructed Long-Necked dinosaur leg and foot bones. Doorway in the background is a standard 6'-6" high [Morrison Museum]

to the foot design of today's largest land animal, the elephant. Elephants walk on their toes and have a large fatty footpad to distribute their large weight on the ground. Long-Necked dinosaurs also grew to be very large and heavy, so it makes sense they would have had a similar foot design to the elephants.

Many biblical creationists, including the author, believe that a Long-Necked dinosaur was the Behemoth described in chapter 40 of the book of Job in the Bible. God's description of Behemoth closely matches our current understanding of the construction of a Long-Necked dinosaur, like *Brachiosaurus*. God's description of Behemoth does not correlate with the comments often offered that Behemoth was an elephant or a hippopotamus. If you doubt this, just ask an uninterested party to describe an elephant or a hippo and see if any of the defining characteristics of these animals match those of Behemoth.

Coelophysis Skull & Skeleton

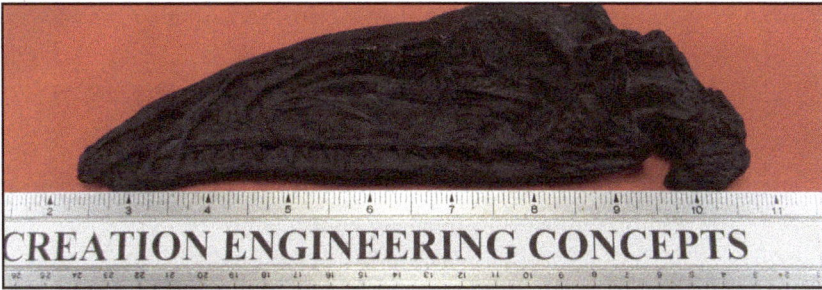

Figure 148: Replica cast of *Coelophysis* dinosaur skull found at Ghost Ranch in New Mexico [DR8].

Description: DR8 skull and cervical segment are about 8 ½" long. Skull is 2 ⅜" high × 1" thick at the top and ⅜" thick at the bottom. The skull was smashed nearly flat prior to fossilization. The largest teeth are ⁵⁄₁₆" long and the eye socket is 1 ¼" diameter.

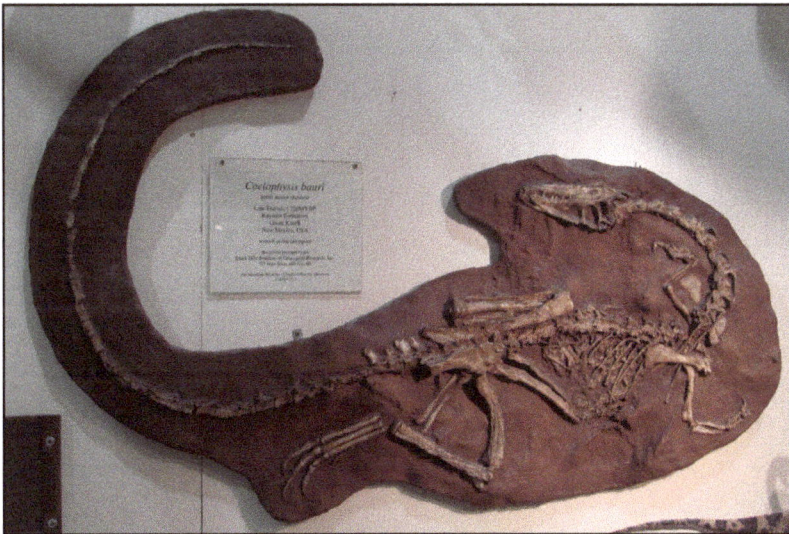

Figure 149: Cast replica of fossil AMNH 7223 *Coelophysis bauri* skeleton from the Kayenta formation of New Mexico at Ghost Ranch [Black Hills Institute].

Interpretation: Relevant "18 Facts" > 1, 2, 3, 16. The dinosaur *Coelophysis* was a Lithe, Fast Running created kind and is mostly known from more than one hundred fossilized skeletons discovered in 1947 at Ghost Ranch in New Mexico. The skeletons are in a bone jumble described by secular scientists as having been caused by a giant flood!

To the biblical creationist the discovered conditions of the large number of dinosaur fossils, some found in the death pose seen in Figure 149, do indeed indicate rapid burial in a giant flood. The condition of the skull DR8 also indicates what one would expect after having been buried in the Genesis Flood at the time of Noah.

Secular science has loosened its grip on some of the uniformitarian ideas once held, but still can never accept the truth of the Genesis Flood because to do so would destroy its atheistic worldview. Geologic catastrophe is now accepted in most secular geology circles, but the great cataclysm carefully described in several chapters of the Bible cannot be accepted due to the hold of naturalism on so many.

Figure 150: Lithe, Fast Running dinosaur artistic reconstruction by Marianne Pike.

Fish Eating Fish

Figure 151: *Euripholis* three-fish fossil plate from Lebanon [MARF33].

Description: Largest fish is *Euripholis boissieri* that is 5 ¾" long and buried eating a small unknown type of fish. Third 1 ¼" long fish at lower right is also an unknown type. The *Euripholis* is commonly called the "Lebanese Viper Fish." Identifying features of this fish are its teeth and characteristic fins and tail.

Figure 152: "Fish aspiration" fossil from the Green River formation of Wyoming [Houston Museum of Natural Science].

Interpretation: Relevant "18 Facts" > 1, 14, 15. The *Euripholis* specimen MARF33 is a "Lagerstatte" (well-preserved) fossil that must have been preserved by rapid and total burial. The burial was so rapid that the one fish was "captured" in the matrix in the process of swallowing the other fish. This is a relatively uncommon fossil finding, yet thousands of such fossils have been found.*

The near perfect "fish eating fish" fossilized display at the Houston Museum of Natural Science shows a *Mioplosus* fish being swallowed by a *Diplomystus* fish. The Houston museum panel explanation reads, "Fish choked to death swallowing another fish." So, according to the secular explanation, these fish choke to death and then slowly sink to the lake floor to be gradually covered by sediment over many years of time. There is nothing about this explanation that matches reality!

Fish *Knightia* and *Diplomystus*

Figure 153: Fish fossils from the Green River formation in Wyoming. Shown are *Knightia* (left) and *Diplomystus* [MARF25, MARF2].

---

* Brown, Walt, *In the Beginning – Compelling Evidence for Creation and the Flood*, 8ᵗʰ Edition, 2008, p. 10.

Description: The *Knightia* fish fossil is 3 ⅝" long × ¹³⁄₁₆" wide and the *Diplomystus* fish fossil is 6 ½" long × 2" wide. The large matrix is 10" × 8." Both are a form of herring.

Interpretation: Relevant "18 Facts" > 1, 2, 3, 8, 9, 10, 14, 17. Untold millions of these fish were rapidly buried in the limestone deposits near Kemmerer, Wyoming by the action of the Genesis Flood. These fossil fish represent some of the more common life forms found in the rock layers of that area.

These fish are just as complex in design and construction as present day herring fish, thus there is no evidence of evolution other than variation within created kinds. There are many similarities between the sedimentary rock layers of Wyoming and those at Solnhofen in Germany where *Archaeopteryx* and many other delicate fossils have been discovered. The Genesis Flood was a worldwide cataclysm as described in the Bible, as attested to by the proper interpretation of these fossils.

Fish *Priscacara*

Figure 154: Fossil fish *Priscacara* from the 18 inch layer in the Green River formation near Kemmerer, Wyoming [MARF65].

Description: The fossil fish is 4 ¾" long × 2 ¾" wide and the matrix is 9 ¾" × 8 ½." Fin and skeletal details of the fish are very visible in this specimen.

Interpretation: Relevant "18 Facts" > 1, 6, 7, 12, 15, 16, 17. Fossil fish *Priscacara* appears like today's perch, and is categorized as such by some experts. Evolutionists believe that the "18 inch" layer from which this fossil was recovered represents about 4,000 years of depositional time. If this is correct, then each year would represent a thickness of only 4 ½ thousandths of an inch, or about the thickness of two human hairs!

Fish and other fossils found in the 18 inch layer have been flattened by the weight of overburden above them, yet still have thicknesses of up to one inch. This *Priscacara* fish fossil has a thickness of at least ⅛ inch. If the depositional rate of 4 ½ thousandths of an inch per year is correct it would have taken 28 years for the deposits to cover the fish. How could the fish remain preserved for even a portion of this length of time? The secular hypothesis is unreasonable. This fish and the other life forms trapped in the 18 inch rock layer were more likely rapidly and completely buried in the worldwide Flood, a watery catastrophe with no equal!

Exploded Fish

Description: The exploded fish fossil consists of most of the skeletal parts of a living fish. The matrix is 6" wide × 6 ½" high × ⅝" thick. The area with the fish remains is about 5" wide × 4" high.

Interpretation: Relevant "18 Facts" > 1, 14, 15. This fossil and others like it are strong evidence that the fish remains were rapidly and completely buried, and supports the biblical creationist worldview. Experiments and observations of dead fish today indicate that they decay within days or weeks at most. In the same Green River formation limestone layers are also found untold numbers of coprolite (fossil poop) samples frozen as fossils. Due to its bacteria content this material would normally decay in a matter of hours.

Figure 155: Fossilized exploded fish from the limestone strata of the Green River formation near Kemmerer, Wyoming [MARF28].

The secular uniformitarian explanations of slow deposition over millions of years for these fossils do not match reality.

Fish *Coelacanth*

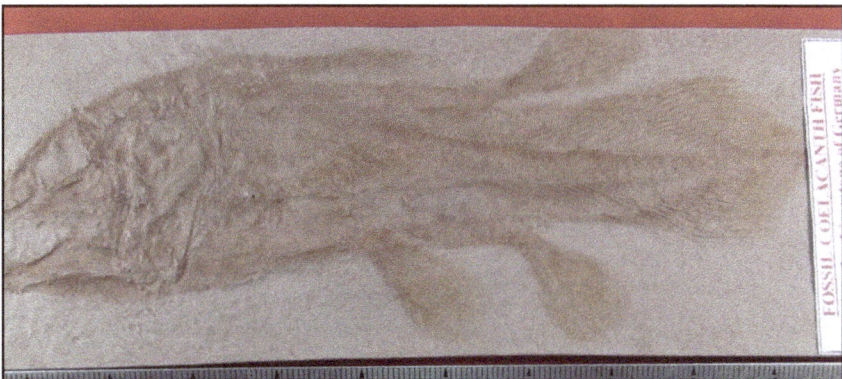

Figure 156: *Coelacanth* replica fish fossil from the Solnhofen limestone of Bavaria, Germany [OR5].

Figure 157: Photo of a model of an extant *Coelacanth* "living fossil" fish [Houston Museum].

Description: The fossil replica *Coelacanth* in Figure 156 on the previous page is 4" wide × 11" long with two fins below and two fins above visible in the matrix.

Interpretation: Relevant "18 Facts" > 1, 6, 14, 15, 16, 17. When the *Coelacanth* was first discovered as a fossil in the rock record it was described by evolutionists as a fish that had lobed fins that were evolving into legs so it could eventually walk on land. Many years later in the 1930s, *Coelacanth* fish were found alive and well in the Indian Ocean. Since then many other of these famous fish have been found alive. The living fish use their fins to swim and there is absolutely no evidence the fins are changing into legs. These fish are another example of a living fossil and are strong supporting evidences for the Genesis Flood and against evolution and millions of years.

Gar Fish *Lepisosteus*

Description: The gar fish is 1 ¾" wide × 9 ¼" long × ³⁄₁₆" thick and the matrix is 7 ½" × 11 ¾".

Interpretation: Relevant "18 Facts" > 1, 6, 15, 16, 17. Gar fish of

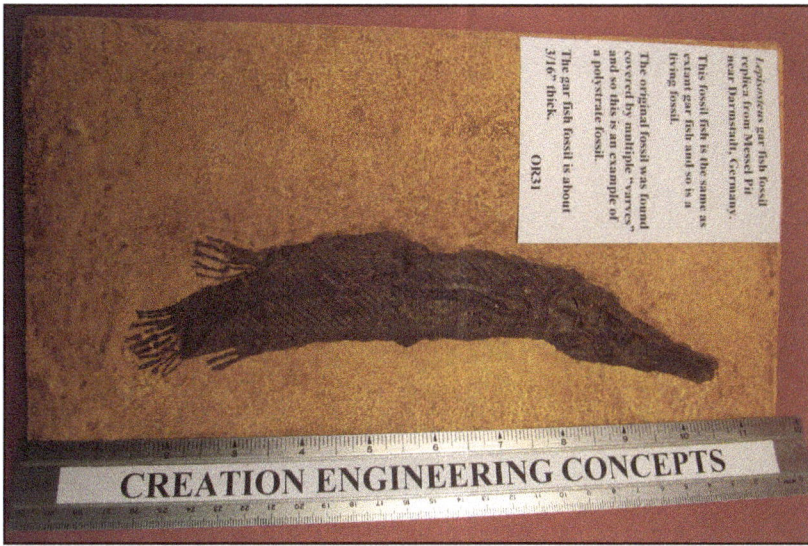

The gar fish fossil is about 3/16" thick.

The original fossil was found covered by multiple "varves," and so this is an example of a polystrate fossil.

This fossil fish is the same as extant gar fish and so is a living fossil.

Lepisosteus gar fish fossil replica from Messel Pit near Darmstadt, Germany.

OR31

CREATION ENGINEERING CONCEPTS

Figure 158: Replica fossil gar fish *Lepisosteus* from the Messel Pit near Darmstadt, Germany [OR31].

today are recognized by their tightly-packed, diamond-shaped scales and their characteristic toothy nose. Fossil gar fish, like the juvenile seen in the figure above, look pretty much the same as the extant fish. There is little change over the evolutionists' imagined fifty million years of progress. Therefore gar fish are another example in the living fossil category that is so often mentioned in this book.

This fossil fish is at least ³/₁₆" thick even after having been flattened under tons of sediments. It was buried by a large number of sedimentary varves (thin layers). That means we can categorize this living fossil as a polystrate fossil as well. Both living and polystrate fossils do not mesh well with the secular worldview of evolution and deep time. The biblical creationist view is shown to be superior time and again by true science.

Onchopristis Sawfish Tooth

Description: The tooth is 2 ½" long and 1" × ⅝" at its root. The

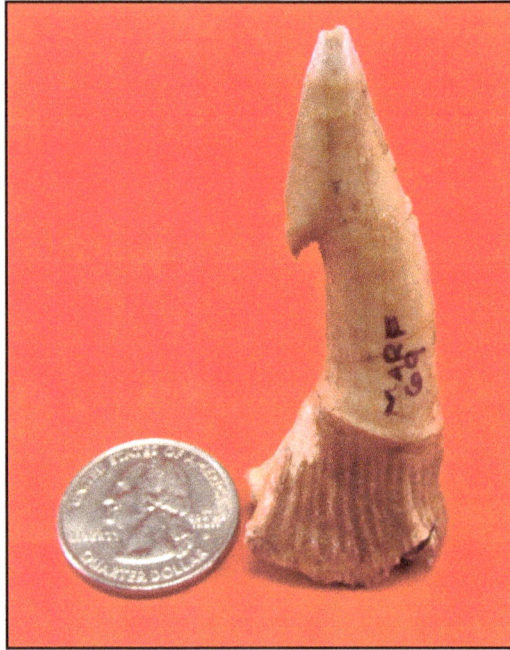

Figure 159: Fossil tooth from the extinct sawfish *Onchopristis* from the Tegana formation of the Kem Kem beds of Morocco [MARF69].

tooth length minus root length is 1 ¾" and the barb is located ⅞" from the pointed end of the tooth.

Interpretation: Relevant "18 Facts" > 8, 10, 14, 17. The *Onchopristis* was a sawfish very similar to the sawfish existing today. The many teeth (denticles) extending from its long rostrum indicate that it would have been a fearsome predator, and it is known to have grown to be over 26 feet long. The fossils found from this fish are usually limited to the barbed teeth and the rostrum since its body was made up of cartilage like rays and sharks, and cartilage does not easily fossilize.

Since modern-day sawfish use the rostrum as their primary sensing mechanism, it is probable that *Onchopristis* had the same ability. The barbs in the teeth that lined the up to 8 feet long rostrum undoubtedly were designed to facilitate retaining any fish impaled on the sawfish's nose.

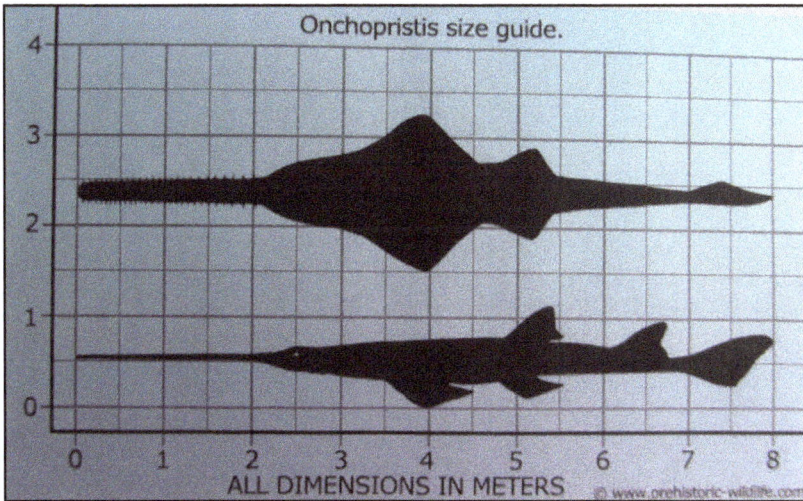

Figure 160: Graph showing large size of extinct *Onchopristis* [credit www.prehistoric-wildlife.com].

Ichthyosaurus Vertebrae

Description: The vertebrae in Figure 161 are ¾" diameter × ½" and ⅝" long. The matrix is 2" wide × 1 ⅜" high × ¾" thick.

Interpretation: Relevant "18 Facts" > 1, 3, 6, 7, 12, 16. This *Ichthyosaurus* fossil is from the same location where 12-year-old Mary Anning (1799 – 1847) discovered the first complete ichthyosaur in England. Anning is a famous storybook character today because of her early fossil collecting in the Lyme Regis area, and because of her subsequent interactions with some of the most famous naturalists of her day.

When Mary Anning discovered her ichthyosaur fossil, the ideas of evolution and millions of years had not yet replaced the biblical explanation for origins within the circles of intelligentsia of England. However, by the time of the publication of Charles Lyell's geology books in the 1830s, and the later publication of *Origin of*

Figure 161: Two connected fossil *Ichthyosaurus* vertebrae in shale matrix from Lyme Regis, Dorset coastline, England [MARF76].

Species by Charles Darwin in 1859, most educated "scientists" had totally rejected the plain reading of God's Word in Genesis chapters 1-11. The impetus behind this monumental change in thinking was the same then as it is today – the attempted rejection of the Creator God, a process succinctly explained by the apostle Paul in Romans chapter one of the Bible.

Figure 162: Photo of complete *Ichthyosaurus* fossil skeleton like that Mary Anning discovered [Houston Museum of Natural Science].

Homology

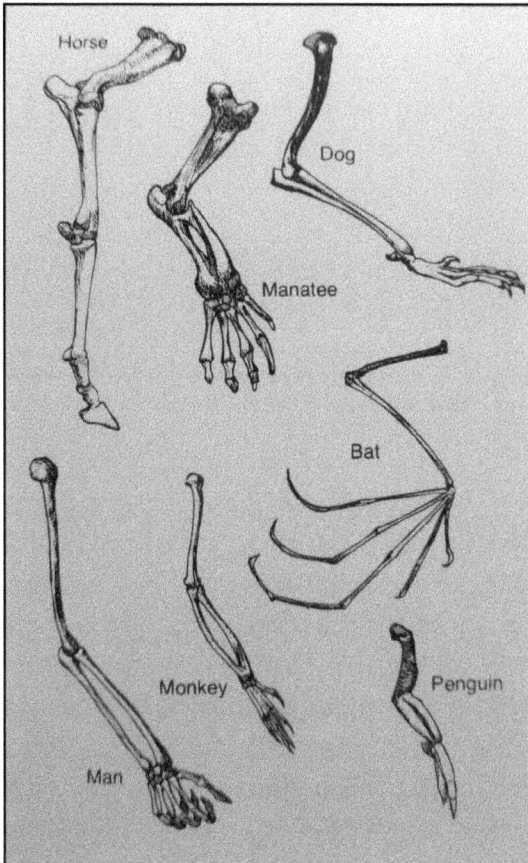

Figure 163: Illustration of the principle of homology that drives evolutionary presuppositional thought.

Ichthyosaurus Flipper

Description: The *Ichthyosaurus* fore flipper in Figure 164 is 10" long ×
3 ¹⁄₈" wide × ⁵⁄₈" thick. The matrix is 6 ⁷⁄₈" × 12 ³⁄₈". There are over
eighty individual bones (tile elements) in the flipper portion of the
fossil. The number of long-axis digits is 5 ½ and the longest digit has
19 tile elements.

Interpretation: Relevant "18 Facts" > 11, 12, 13, 14, 18. One of the

Figure 164: Fossil replica *Ichthyosaurus* fore flipper from a specimen from Somerset, United Kingdom [OR27].

presuppositions of the evolutionary worldview is that homology proves evolution. That is, since different kinds of animals have similar morphology or design, then they must have a common ancestor from which they all evolved. Evolutionists teach that five-fingered hands (manus) all evolved from the "early tetrapods." The proposed evolution of whales from land animals is largely predicated on the existence of five fingers in the whale forelimbs.

It is interesting to note that the marine animal currently being considered, the *Ichthyosaurus,* has paddles that at first glance look similar to those of whales. But the *Ichthyosaurus* did not have a "hand" with the familiar five fingers with separately-connected finger bones as do the whale, the manatee, the monkey, and man (see Figure 163). Instead it had tile-shaped bones arranged on the long axis as digits. In addition, the ichthyosaur digits are all squashed tightly together into a solid functional paddle-like appendage. Fossil ichthyosaurs have been found with from three to eight digits and up to thirty elements per digit.

The evolutionist believes that the ichthyosaurs "developed" this odd design for its flippers all on its own. The biblical creationist believes that God was the designer that provided this unique and functional way to allow the creature to navigate in the water. Nevertheless, since the ichthyosaurs are thought to be extinct today it is possible that there were inherent disadvantages to this flipper design which contributed to the demise of the ichthyosaurs during or after the Flood.

Mosasaurs and their teeth

Figure 165: Fossil teeth from three species of mosasaurs (l to r) all from Morocco: *Platecarpus ptychodon, Tylosaurus* sp with root and *Mosasaurus beaugei* [MARF21, 29, 22].

Description: The *Platecarpus ptychodon* tooth is 1 ⅛" long × ½" wide at the base and has noticeable longitudinal grooves starting halfway up from the base. The *Tylosaurus* sp tooth is 1 ½" long × 1 ¼" wide but does not have any noticeable grooves. Including the root it is 4 ½" long. The *Mosasaurus beaugei* tooth is 1 ½" long × ¾" wide without any noticeable grooves.

Interpretation: Relevant "18 Facts" > 8, 9, 10, 14. As with most life kinds considered to be extinct, as are the mosasaurs, scientists have assigned numerous genera and species names to the fossils of these marine animals. There are at least a dozen mosasaur species, but some may be the same marine animal exhibiting variation within a created kind. The author has noticed that some mosasaur, pliosaur

Figure 166: Close-up look at *Tylosaurus* teeth in mandible [Morrison Natural History Museum].

and plesiosaur teeth have longitudinal grooves on the exterior surfaces and some don't, as seen in the figure above. It is not possible for anyone to know for sure whether the existence of these tooth grooves is an indicator of a created kind differentiation. This issue is always a problem for both the biblical and evolutionary views.

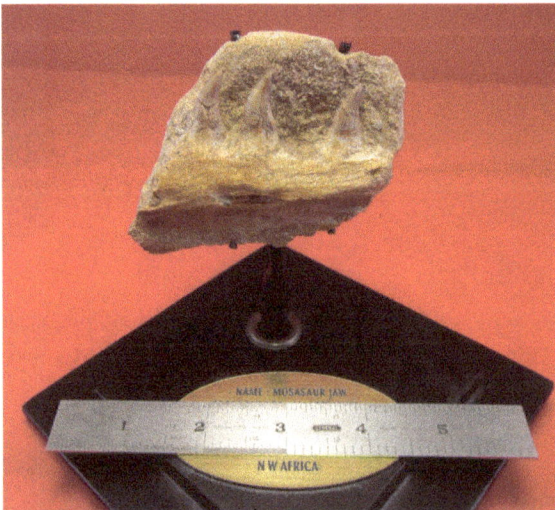

Figure 167: Fossil *Mosasaurus* jaw fragment with three teeth in matrix from a phosphate deposit at Khouribga, Morocco in Africa [MARF73].

Description: The mandible section is 2 ⅞" long × ⅞" high × ⅜" thick. The three teeth are all about ⅞" long × ⁷⁄₁₆" wide at the base × ¼" thick at the base. The matrix measures 3" wide × 2 ⅜" high × ¾" thick.

Interpretation: Relevant "18 Facts" > 1, 2. Ever since the discovery of *Archaeopteryx* and the publishing of *The Origin of Species* by Charles Darwin in the mid-nineteenth century, fossil fraud has been a big industry. This is because certain fossils have become extremely valuable if they can be imagined to be a transition from one kind to another kind, that is, be described as a "missing link." Other fossils are popular because they are familiar and are relatively easy to fake. Many fossils from Morocco are not authentic because they are modified using skillful combinations of fossil fragments glued to look like something they are not. This is true of some Moroccan mosasaur fossils and also some of the rarer trilobite fossils.

The *Mosasaurus* fossil in Figure 167 from Morocco is valuable because it is a piece of bone with three attached teeth connected

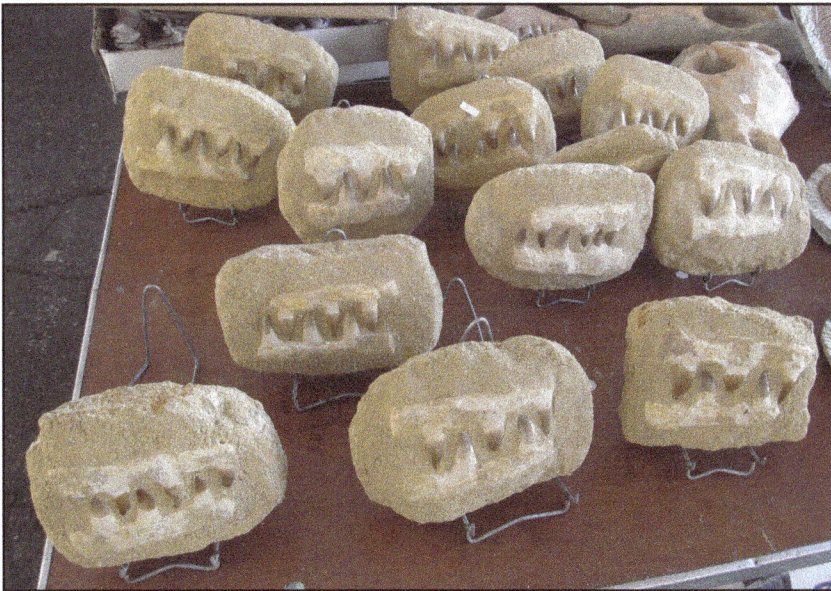

Figure 168: Moroccan products for sale showing mosasaur teeth inserted into a matrix. It is doubtful that any of these are authentic fossils.

to the rock matrix in which they were all found. That is, the fossil is displayed as it was found *in situ,* and is not an artistic artificial combination of fossil, rock and glue that has been designed and constructed by forgers.

The fossils in Figure 168 are typical of those that Moroccan workers have modified to be something they are not. Sometimes, it is very difficult for the layman and even experts to tell the difference between authentic and non-authentic fossils.

The practice of non-authentic fossil manufacture is not restricted to Morocco. It has been a huge problem in Europe as addressed previously regarding *Archaeopteryx* and is an identified problem for some fossils from China as well.

Figure 169: *Mosasaurus conodon* reconstruction from the South Dakota Pierre Shale [South Dakota Museum of Geology].

Mosasaur *Globidens*

Description: The tooth to the left in Figure 170 is 1" diameter × ⅞" long. The rooted tooth is 1" diameter (root is 1 ¼" diameter maximum) × 3 ⅛" overall length. The sign in the center shows a photo of a fossil *Globidens* tooth-filled lower jaw.

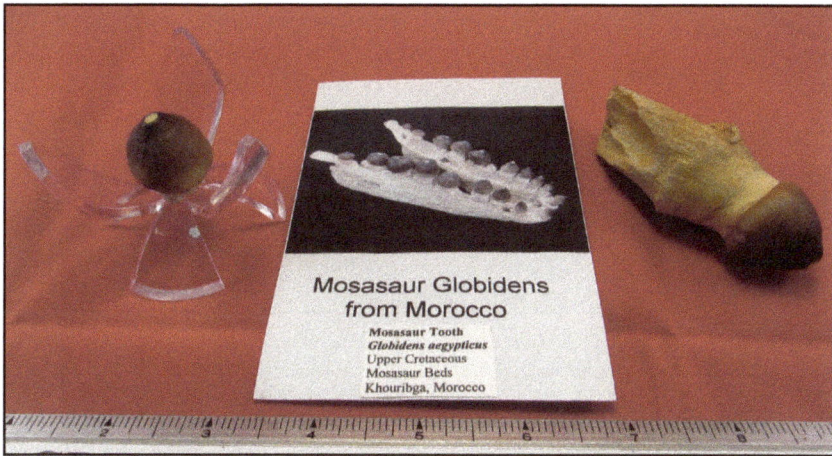

Figure 170: Fossil teeth from the mosasaur *Globidens aegypticus* from Khouribga, Morocco. Tooth only at the left, tooth with root at right [MARF20, 30].

Interpretation: Relevant "18 Facts" > 8, 9, 10, 13, 14. Most mosasaurs are thought to have been very large marine animals up to 60 feet long. And they all had very large, sharp teeth as seen in the five earlier figures of this book. But, *Globidens* was a mosasaur that had teeth with an entirely different design. Rather than being pointed on the ends like other mosasaurs, *Globidens* teeth were similar to rounded nubbins on the ends. Secular scientists believe that *Globidens* developed this type of teeth to be better able to use clams and other sea animals with protective shells as food. They would propose that these teeth evolved from other tooth designs. However, no intermediate designs have yet been discovered in the rock record.

A biblical creationist explanation for the teeth of *Globidens* would be that, in the beginning, God created this kind of animal with rounded teeth for His own purposes.

Pliosaurus Tooth

Description: The replica *Pliosaurus* tooth is 9 ½" long overall and the tooth itself has a mostly triangular cross-section. About

Figure 171: Short-necked marine reptile *Pliosaurus* replica tooth fossil [OR19].

one-third of the tooth surface starting at the pointed end has longitudinal grooves. This specimen has a root about 6" long and the root cross-section is 2" wide × 1" thick.

Interpretation: Relevant "18 Facts" > 13, 14. Pliosaur and plesiosaur reconstructed fossil skeletons are similar in that they both have four paddles and a relatively short tail. However, pliosaurs had short, thick necks and plesiosaurs had very long slender necks. *Pliosaurus* was 33 to 40 feet long with large powerful jaws resembling those of a crocodile. Plesiosaurs had sharp teeth but not the powerful jaws or the large head of the pliosaurs. Usually if animals have large differences in the teeth design, like is seen here between pliosaurs and plesiosaurs, they would not be considered close relatives, but evolutionists believe these two genera have a common ancestor.

Because of the large tooth and front-end differences the author believes it is possible that these two animals are different created kinds.

Plesiosaur Teeth

Description: The plesiosaur tooth in Figure 173 from France is ³⁄₈" long × ¹⁄₈" maximum diameter and has full-length longitudinal striations. It is not possible to know if this is a complete tooth or just the tip of a tooth. The tooth is jet black in color.

Figure 172: *Plesiosaurus* skeletal reconstruction showing long neck and smaller teeth that differ from pliosaurs [Houston Museum of Natural Science].

Figure 173: Fossil plesiosaur tooth in matrix from Nancy, France (left), fossil plesiosaur tooth from the phosphate beds of Morocco (center), fossil *Nothosaurus* tooth in matrix from Rothenberg, Bavaria (right) [MARF6, 19, 23].

The Moroccan plesiosaur tooth is 2 ⅛" long × ½" maximum diameter and has no striations. The tooth is dark brown in color.

The *Nothosaurus* tooth is 1" long × ¼" maximum diameter and has longitudinal striations for the full length of the tooth. It is missing about ¼" from the tip end. The tooth is jet black in color.

Interpretation: Relevant "18 Facts" > 6, 7, 8, 9, 16. Identified fossil plesiosaur teeth that are available from fossil dealers worldwide are either black or dark brown in color. Some have striations and some do not. There is not any noticeable difference in the color and structure of the two black teeth in the above figure, so why is one identified as a *Plesiosaurus* tooth and the other as a *Nothosaurus* tooth? The author believes this is because one was found in rock assumed to be millions of years older than the other. Therefore, according to evolutionary presuppositions, they cannot be the same animal and must be separated by millions of years of evolutionary change and extinction.

With the biblical presupposition of a one-year-long worldwide Flood having buried both teeth, it is possible to conclude that both teeth came from the same created kind of marine animal.

Super Crocodiles

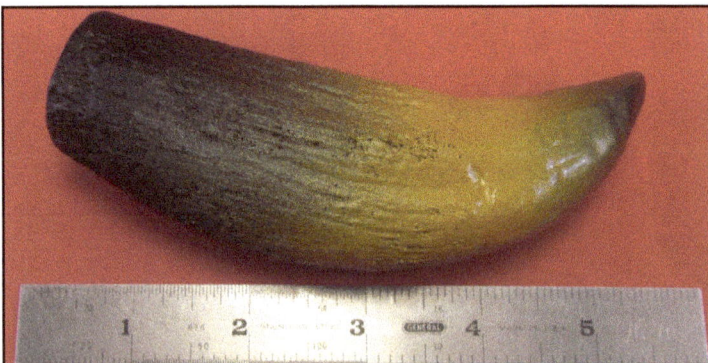

Figure 174: Super crocodile replica fossil tooth from *Sarcosuchus* [OR20].

Description: The *Sarcosuchus* replica tooth is 1 ¾" wide × 1 ¼" thick × 6" long. It has an oval cross-section except for the tip.

Interpretation: Relevant "18 Facts" > 6, 8, 9, 10, 17. The jaws of existing crocodiles have a variety of tooth sizes in the upper and lower jaws and it is difficult to determine where in the jaws a particular fossil tooth originated. The largest extant crocodile tooth the author has seen is about the size of his pinkie finger, so the *Sarcosuchus* tooth shown above is very impressive.

Today's saltwater crocodiles can grow to be over twenty feet long and weigh a ton. *Sarcosuchus* super crocodiles have been reconstructed to be nearly forty feet long with an expected weight of many tons. Evolutionists agree that crocodiles are living fossils and are not much different from those found in the rock record. This means that the biblical creationist view that crocodiles were created only thousands of years ago makes sense, and the various super crocodiles like *Sarcosuchus* and *Deinosuchus* from the rock record are likely the same created kind as are today's saltwater crocodiles. The author believes that a super crocodile was the Leviathan that God describes in chapter 41 of the book of Job in the Bible.

Figure 175: Super crocodile *Deinosuchus* skull [Brigham Young University Museum].

Figure 176: "Living fossil" display of crocodiles where fossil and extant crocodile skeletons are identical [Smithsonian Museum].

Sarcosuchus Armor Plate

Figure 177: *Sarcosuchus* super crocodile replica fossil armor plate (scute) [OR25].

Description: The replica scute (osteoderm) in Figure 177 is 5 ½' wide × 6 ⅛" long × 1 ½" thick. The exterior surface is shown in the photo and the inside surface is relatively smooth. The exterior ridges are up to ⅝" high.

Interpretation: Relevant "18 Facts" > 1, 10, 17. If *Sarcosuchus* was indeed Leviathan then its armor plate as seen above could certainly have been impenetrable by men as stated by God in chapter 41 of the book of Job.

Crocodile Scutes

Figure 178: Twenty fossil scutes (osteoderms) of order Crocodilia animals from the Hell Creek formation in Carter County, Montana [MARF75].

Description: The fossilized osteoderms range in size from 1 ⅝" × 1 ⅜" × ³⁄₁₆" thick to ⅞" × ⅞" × ³⁄₁₆" thick. Most of the osteoderms are slightly curved on the backside and the remainder are basically flat. The nine specimens to the right of the photo tend to have raised medial ridges while the remaining eleven do not have this characteristic.

Interpretation: Relevant "18 Facts" > 8, 12, 14, 17. The Hell Creek formation covers portions of Montana, North Dakota, South Dakota and Wyoming. Intense studies over the past 150 years have found many fossil species of almost every kind of life imaginable in this rock formation. Many dinosaur fossils have been discovered there as well as the crocodilians represented by the osteoderms shown above.

Experts differ in their descriptions of the extent and design of the surface scutes and the underlying bony osteoderms in the various living crocodilians. There is also little consensus regarding what type of osteoderm matches crocodiles versus alligators in the fossils. Some experts say fossil osteoderms with medial ridges are from alligators and those without are from crocodiles. However, the super crocodile *Sarcosuchus* had ridges on its armor plates.

One thing that is sure from the rock record is that crocodiles and alligators similar to those living today existed at the same time as the monster crocodiles, like *Sarcosuchus,* prior to the global Flood. We know that because we find their teeth, osteoderms and other hard parts in the rocks remaining from the catastrophe at the time of Noah.

Fossil Bird *Gallinuloides*

Figure 179: Replica cast of quail-like bird *Gallinuloides* from the limestone of the Green River formation in Lincoln County, Wyoming [OR10].

Description: This *Gallinuloides* bird skeleton is 4 ⅛" tall with a skull 1 ½" long. The matrix is 6" × 6."

Interpretation: Relevant "18 Facts" > 1, 2, 3, 15, 17. While millions of fish are fossilized in the limestone layers of the Green River formation, birds are very rare. This is to be expected since they would have been able to fly above the worldwide Flood commotion below them and were less likely to be quickly buried. Many birds would have been able to exist for awhile by landing on mats of floating vegetable matter before eventually dying of starvation. Their remains would have rotted away in the same way as the remains of dead birds do today.

Pterodactylus Skeletons

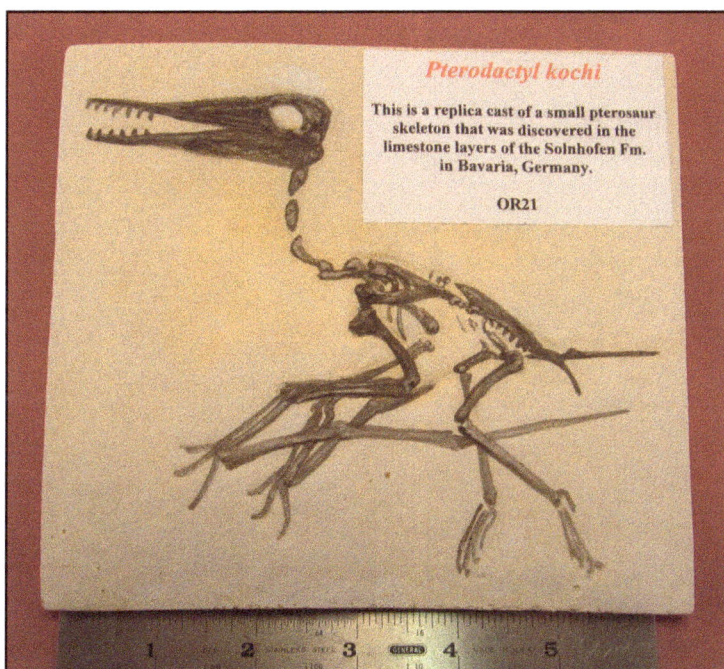

Figure 180: Replica cast of *Pterodactylus kochi* complete skeleton from the limestone layers of the Solnhofen formation in Bavaria, Germany [OR21].

Figure 181: Replica cast of *Pterodactylus antiquus* complete skeleton from the limestone layers of Solnhofen [OR23].

Description: *P. kochi* dimensions are 2 ½" skull length, $^{15}/_{16}$" femur length, $^{15}/_{16}$" humerus length, and matrix size of 6 ⅜" wide × 5 $^{9}/_{16}$" high.

P. antiquus dimensions are 3 ⅛" skull length, 1 ⅛" femur length, 1 ⅛" humerus length, and matrix size of 6 ⅞" wide × 8 ¼" high.

Interpretation: Relevant "18 Facts" > 8, 9, 10. Numerous pterosaur specimens identified as genus *Pterodactylus* have been discovered, especially from Solnhofen sediments. Some evolutionists believe that only *Pterodactylus antiquus* is a valid name and *P. kochi* should be incorporated into the *P. antiquus* classification.

The two specimens shown above are very similar, but *P. antiquus* is slightly larger. This could be attributable to sex or age differences. Any minute differences in skeletal design could be due to variation within the kind that accumulated during the 1,600 years from the time of the fall to the time of the Genesis Flood.

Rhamphorhynchus Skeleton

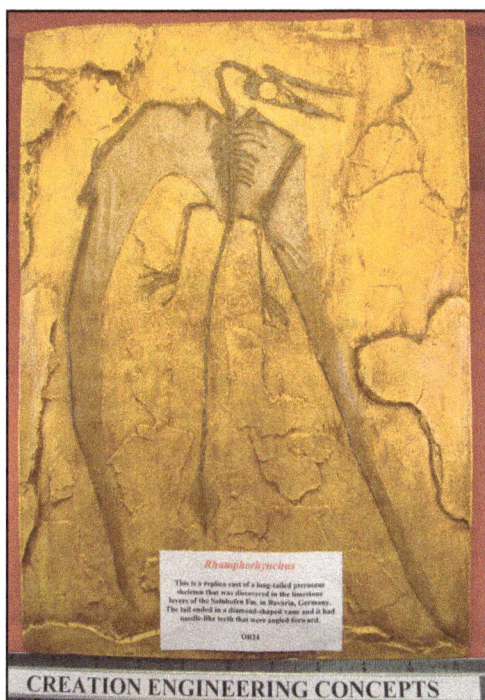

Figure 182: Replica cast of the complete skeleton of the long-tailed pterosaur *Rhamphorhynchus* from Solnhofen in Bavaria, Germany [OR24].

Description: Pterosaur dimensions are 2 ⁷⁄₁₆" skull length, ⁹⁄₁₆" femur length, ¾" humerus length, 6 ¾" tail length, and matrix size is 10" wide × 13 ¾" high.

Interpretation: Relevant "18 Facts" > 9, 14. Pterosaur *Rhamphorhynchus* is extremely different in design from *Pterodactylus* as well as the well-known *Pteranodon.* These three pterosaurs are all different kinds of flying reptiles from the rock record that were created on day five, and then 1,600 years later, rapidly buried in the worldwide Flood in Germany and elsewhere around the world.

Many cryptozoologists believe that recent sightings around the world indicate that some kinds of pterosaurs still exist. If so, these would be additional examples of living fossils that would add to the huge body of evidences that discredit evolutionary hypotheses.

Figure 183: Museum reconstruction of Pterosaur *Rhamphorhynchus* from Germany [Wyoming Dinosaur Center].

Trionyx Turtle

Description: The matrix in Figure 184 is 9" wide × 12 ½" high and the skull is 2 ¼" long. The turtle shell is 4 ⅜" wide × 5 ⅛" long and has ⅜" and ½" diameter indentations 1 ⅛" apart that evolutionary scientists suggest are bite marks from an alligator or crocodile.

Interpretation: Relevant "18 Facts" > 1, 6, 15, 16, 17. The secular explanation for the millions of fish and other fossils (including this turtle) discovered in the limestone quarries of the Green River formation is a slow covering of 4 ½ thousandths of an inch of sediment per year over a period of 4,000 years. This turtle fossil is flattened from the overburden, but is still about ¼" thick. So, according to the uniformitarian view, it would have taken at least sixty years to cover the turtle during the fossilization process. Common sense tells us this is not a reasonable explanation and a catastrophic cause makes far more sense.

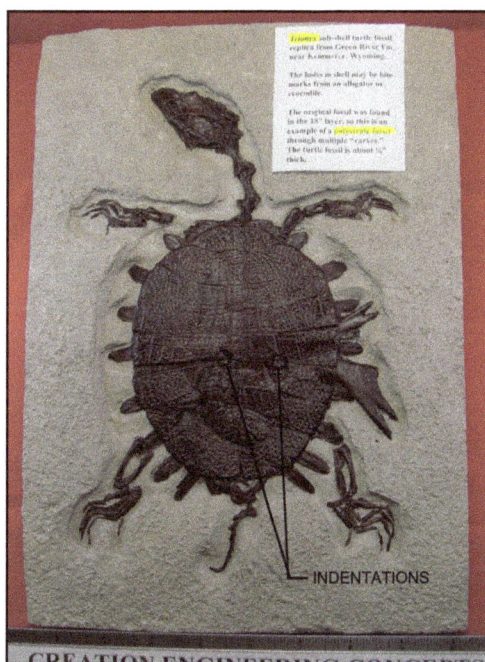

INDENTATIONS

Figure 184: *Trionyx* soft-shell turtle fossil replica cast from the Green River formation near Kemmerer, Wyoming [OR18].

The author personally inspected the actual fossil *Trionyx* from which this cast was made. While it is possible that the indentations are from the teeth of a crocodilian, the entire rear portion of the shell is depressed, leading the author to be open to other causes for the indentations. Those possibilities are nearly endless if it is understood that the fossilization process for this turtle was cataclysmic in origin.

Trionyx is a living fossil that is still abundant today in lakes, estuaries, and slow-moving rivers. And this specimen is also a polystrate fossil. Living fossils as well as polystrate fossils can be used to effectively argue against deep time and for support of the biblical timeline.

Propelodytes Frog

Description: The fossil frog is 2 ⁷/₈" wide × 5 ³/₈" long and its skull is

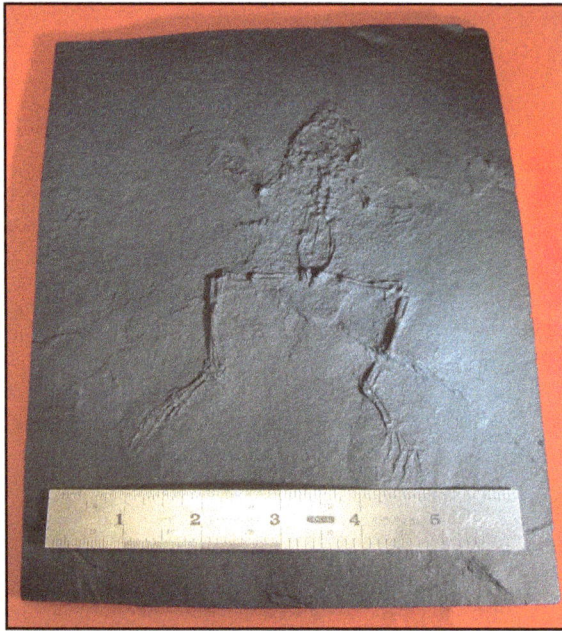

Figure 185: Fossil replica *Propelodytes* frog in shale matrix from the Messel Pit near Darmstadt, Germany [OR29].

1 ¹⁄₈" wide × 1" long. The matrix is 6 ³⁄₄" wide × 8 ¹⁄₈" long. The replica matrix color shade is somewhat darker than the actual shale matrix.

Interpretation: Relevant "18 Facts" > 1, 6, 14, 17. This *Propelodytes* fossil frog has the slender streamlined body and the pointed head of frogs of today. Its large feet are attached to powerful hind legs that provide the frog with the thrust needed for it amphibious environment. This fossil frog is dated at fifty million years old by evolutionists, yet it looks the same as frogs of today. There are varieties of frogs and some extinct frogs, but they all are frogs with none of them seen evolving from, or to, some other life form.

Many other kinds of animals and plant fossils have been found in the Messel Pit. They are all complete and functional with no indication of morphological transition into something else. The Messel fossils are split from the shale in a manner similar to the way fossils are recovered from the Solnhofen limestone of

Germany, the Green River limestone of Wyoming and other sites around the world.

Eryops Amphibian

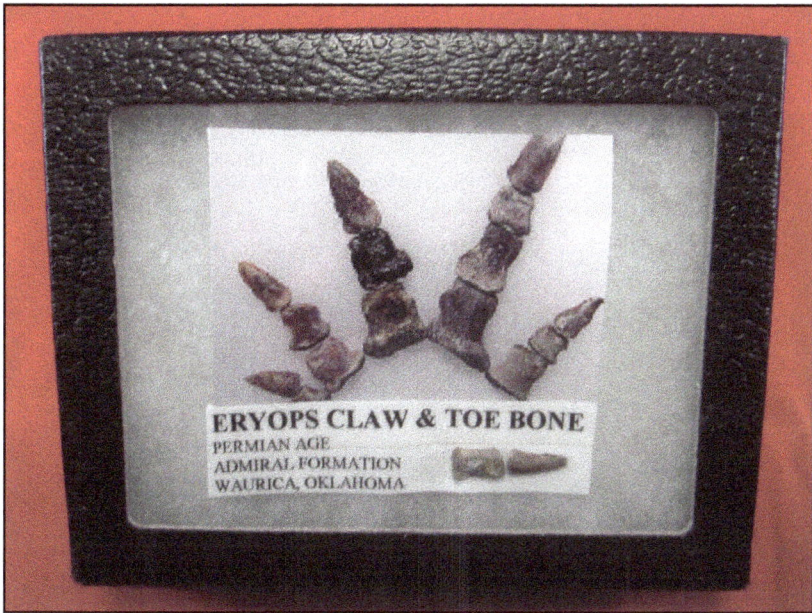

Figure 186: Extinct amphibian *Eryops* fossil claw and toe bone (lower right of figure) superposed on a photo of the bones of a complete *Eryops* foot. The two fossil pieces are from the Admiral formation near Waurica, Oklahoma [OF11].

Description: The Riker display box is 4 ⅜" x 3 ⅜" and the claw is ⅜" long. The toe bone is ⁵⁄₁₆" long × ³⁄₁₆" diameter.

Interpretation: Relevant "18 Facts" > 8, 9, 10, 14, 18. *Eryops* was an amphibian with a skeletal design similar to frogs, toads and salamanders. However, fossils of *Eryops* have been found that indicate it could grow to be five feet long with a large triangular skull and short, powerful legs. Amphibians lay their eggs in water and cannot live long in areas without standing water.

Many mammals, reptiles, amphibians, birds and dinosaurs have

Figure 187: Museum reconstruction of *Eryops* skeleton [Royal Tyrrell Museum].

similar designs for their arms and feet. That is, these component parts are homologous. Evolutionists explain these similarities as the result of evolution from a common ancestor over millions of years. To the evolutionist homology is the conclusive proof of evolution. The biblical creationist view is that God used similar designs at the time of creation for similar functions in different animal kinds. While variation within kinds is allowed by the DNA in creatures, there is no macroevolution of one kind into another kind.

Icaronycteris Bat

Description: The replica fossil bat skeleton is complete. The bat is 4 ⅝" long × 2 ⅜" wide and the matrix is 2 ⅝" × 5 ¼".

Interpretation: Relevant "18 Facts" > 1, 10, 17. This particular fossil holotype for this kind of bat was discovered in the early 1930s and

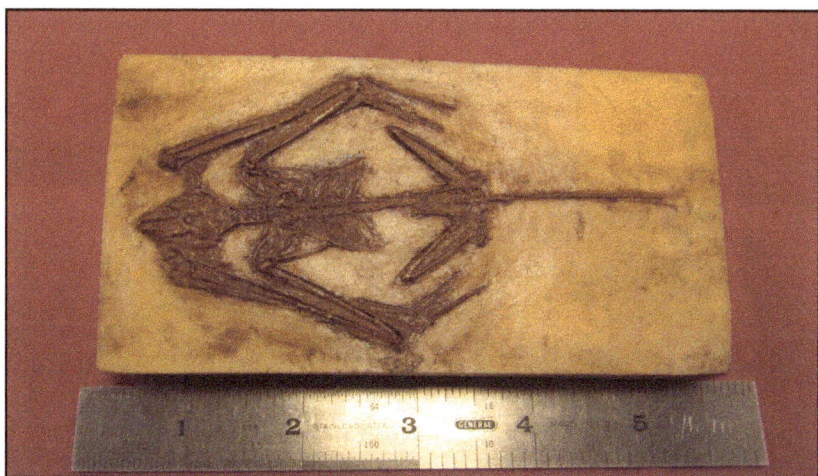

Figure 188: Replica fossil bat *Icaronycteris* from the Green River formation near Kemmerer, Wyoming [OR30].

is among the "oldest" bat fossils discovered so far. With the genus name *Icaronycteris* it looks like it could be a "mouse-tailed" type of bat like those in existence today.

As with all mammals, evolutionists cannot concoct a family tree for bats from the beginning of their supposed evolution, and so have no idea what the bat's common ancestor might be. The author believes that there are lots of genera of bats living today that are variations of the original kinds of bats that God created in the beginning.

What the rock record shows is bats with bat characteristics, and not animals developing bat characteristics from other animal kinds. Bats are bats and always have been bats. That sounds like a description of creation, not evolution!

Oreodont *Merycoidodon*

Description: The image in Figure 189 is one documenting the preparation of MF7 by the author in September, 2012. The completed preparation is shown in Figure 190.

Interpretation: Relevant "18 Facts" > 1, 2, 3, 6, 7, 15. The author

Figure 189: Oreodont partial fossil skull *Merycoidodon* as removed from the sediments of the White River formation of South Dakota [MF7].

has some experience with mammal fossils from the White River formation of South Dakota and Nebraska as well as from the John Day Fossil Beds of Oregon. The mammal fossils of dogs, cats, tapirs, rhinos, horses, camels, oreodonts, peccaries, and entelodonts have similar morphology at these two American badland locations some 800 miles apart. An in depth exposition on these fossils can be seen in the author's book, *Discovering the Animals of Ancient Oregon.*

In most cases the recovery of a mammal fossil consists of removing fragmentary partial bones and teeth. Removing the bone parts from the matrix can be a tedious and time consuming process. All of these badlands fossils were obviously rapidly buried and possibly transported due to a watery catastrophe. Special glues are required to piece the fossil parts together in some semblance of their original construction.

Figure 190: Oreodont mammal partial fossil skull *Merycoidodon* [MF7] as prepared by the author and *Merycoidodon* jaw and tooth section [MF5]. Both specimens were recovered from the White River formation of South Dakota.

Description: The *Merycoidodon* partial skull is 5" wide × 4 ¼" deep × 8" long. Most of the teeth are still in the skull and mandible, but the nose is missing. The [MF5] jaw section contains 3 ½ teeth and is 1 ½" wide × 1 ½" deep × 2 ⅛" long.

Interpretation: Relevant "18 Facts" > 1, 2, 3, 6, 7, 15. Oreodonts are extinct animals that are found in the rock record as a result of the global Flood at the time of Noah. Fossils indicate there were two distinct kinds, one with hoofs and the other with clawed feet. Why and when these animals went extinct is a mystery, but we know they were around for at least the 1,600 years after the fall because they are found in the rock record.

Archaeotherium Entelodont

Figure 191: Two fossil upper molar teeth in a jaw section of *Archaeotherium* from the White River Badlands, Pennington County, South Dakota [MF12].

Description: The teeth are about ¾" long × ¾" wide × 1" deep. The jaw section is 1 ¼" wide × 1 ⅜" high × 2 ¼" long.

Interpretation: Relevant "18 Facts" > 1, 2, 3, 14, 15. The *Archaeotherium* is an extinct entelodont known from the White River Badlands and the John Day Fossil Beds of Oregon as well as locations in Europe. While a few skulls and skeletal bones of these giant pigs have been discovered, the most common fossils found are teeth like those in the figure above.

Due to the large, up to three feet long, skulls with ferocious looking teeth of *Archaeotherium*, reconstructions of these animals are commonly showcased at natural history museums throughout the world. The protuberances on the faces of these animals look similar to those on warthogs of today.

Figure 192: Replica skull of extinct *Archaeotherium* from the Brule formation, White River Badlands of South Dakota [Royal Tyrrell Museum].

The biblical creationist and secular evolutionist views of extinction are very different. According to the biblical view, everything was created by God "very good" in the beginning. There was no death, suffering, disease or extinction until Adam sinned and brought death into the world. Many of the life forms that have gone extinct did so as direct and consequential results of the worldwide Flood. The Bible promises that these aspects of the world will continue until Jesus Christ destroys the curse and brings a new heaven and a new earth.

The evolutionist believes that death, disease, suffering and extinction are necessary component parts of the advances over millions of years that have brought the progress that has resulted in the advent of man on earth. To the evolutionist, extinction and death are necessary parts of evolution; and the end game for the evolutionist is heat death of the entire universe.

Camel *Poebrotherium*

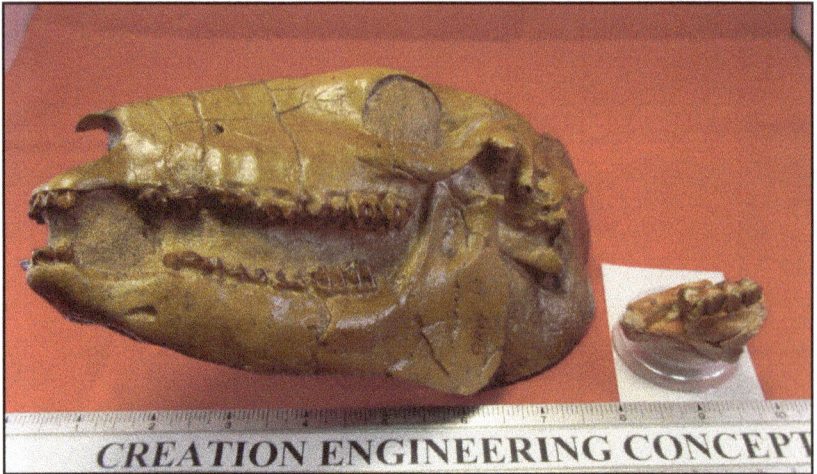

Figure 193: Fossil replica skull of *Poebrotherium* a camel found in the White River Badlands of Nebraska [OR17] and fossil teeth of *Poebrotherium* found in the White River Badlands of South Dakota [MF10].

Description: The replica *Poebrotherium* camel skull is 8" long × 3 ⅞" high × 1 ⅞" wide maximum. The skull is nearly complete and relatively undamaged. The fossil camel teeth section in matrix consists of four molar teeth probably from the mandible. The teeth/bone section is 1 ½" long × 1 ¼" high × 1" wide maximum.

Interpretation: Relevant "18 Facts" > 1, 2, 3, 7. Many of the mammal fossils found in the White River Badlands are very similar to those found in the John Day Fossil Beds of Oregon. One factor both locations have in common is that the mammal fossils recovered are usually fragmentary and disarticulated. Seldom is a complete camel skull like the one in the figure above found, so the fossil fragments must be identified by the characteristics of the teeth.

The condition of the camel fossils discovered at these locations can best be explained by tectonic, volcanic and hydraulic forces assumed by biblical creationists to have acted during the Genesis Flood at the time of Noah.

Mesohippus Skull

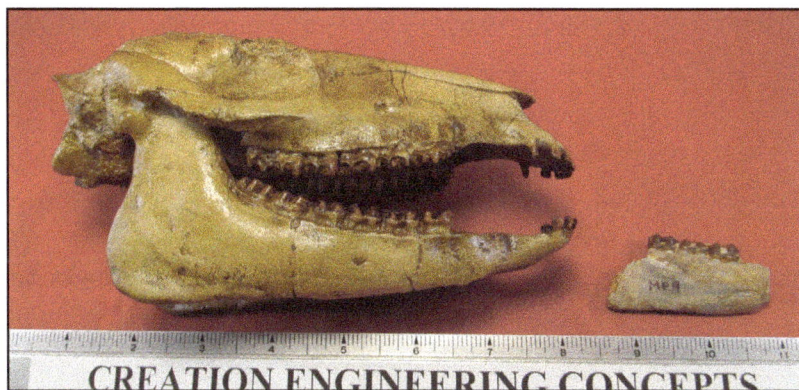

Figure 194: Fossil replica skull of *Mesohippus*, a 3-toed "horse" found in the White River Badlands of Nebraska (left) [OR12]. Also fossil teeth of *Mesohippus* found in the White River Badlands of South Dakota (right) [MF9].

Description: The replica skull is 8" long × 4" high × 3" wide maximum. The skull is complete and relatively undamaged. The fossil teeth section consists of three complete molar teeth from the mandible. The tooth and bone section is 2 $\frac{3}{8}$" long × 1 $\frac{1}{4}$" high × $\frac{3}{8}$" wide.

Interpretation: Relevant "18 Facts: > 1, 2, 3, 7, 9, 10, 12, 13. Evolutionist presuppositions start with a belief in a "common ancestor" and so evolutionists must string together fossils in a manner leading back toward said beginning life form. Biblical creationists start with a different set of presuppositions, including that God created according to kinds in the beginning. While all agree there is variation within a kind, the two paradigms are far apart when it comes to how much variation has occurred and can possibly occur.

The author believes that in the so-called "horse series" *Mesohippus* was created within the family including genera *Hyracotherium*, *Orohippus*, *Miohippus* and *Protohippus*. *Pliohippus* and *Equus* were created within a separate created kind.*

---

* For details see: Mitchell, J.D., *Discovering the Animals of Ancient Oregon*, Leafcutter Press, 2013, pp. 172-183.

Peccary *Tayassu*

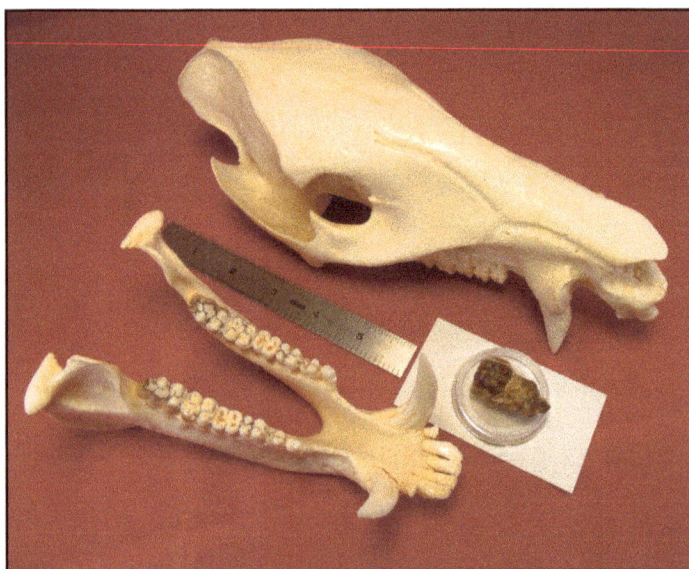

Figure 195: Actual *Tayassu tajacu* peccary mandible and skull [OR15]. Also fossil teeth of *Perchoerus* found in the White River Badlands of South Dakota (center-right) [MF8].

Description: The actual peccary mandible is 6 ¾" long × 3" high × 3 ¾" wide maximum. The actual peccary skull is 10" long × 3 ⅜" high × 4" wide maximum. All the teeth on the *Tayassu* mandible and skull show very little wear. The *Perchoerus* tooth/bone section consists of 2 ½ molar teeth and is 1 ½" long × ¹⁵⁄₁₆" high × ¾" wide. The molar teeth of *Perchoerus* show some wear and are about thirty per cent larger than the corresponding *Tayassu* teeth.

Interpretation: Relevant "18 Facts' > 8, 9, 10, 13, 17, 18. Peccaries are known as javelinas in the United States and are found wild in southern Arizona, New Mexico and Texas. They also inhabit areas of South America. They are not pigs, but have a number of similarities to pigs. In Figure 195, the skulls and teeth of the *Tayassu* specimen and the fossil *Perchoerus* are very similar. The molar teeth of peccaries and humans are very similar as well. They are so similar that an extinct peccary fossil tooth was once misidentified by evolutionists as belonging to an ape-like human ancestor. The imagined human ancestor with the peccary-like teeth was called Nebraska Man.* When all the characteristics of peccaries are considered, an argument can be made that it makes sense to categorize the peccary as another living fossil.

In the author's book, *Discovering the Animals of Ancient Oregon*, he wrote, "The facts are harmonious with the biblical creationist view that peccaries are one created kind, the variations found are due to God-designed built-in genome adaptability prior to and after the Flood, and their damaged skulls and teeth are found scattered about as a result of the Genesis Flood."**

Hoplophoneus Cat

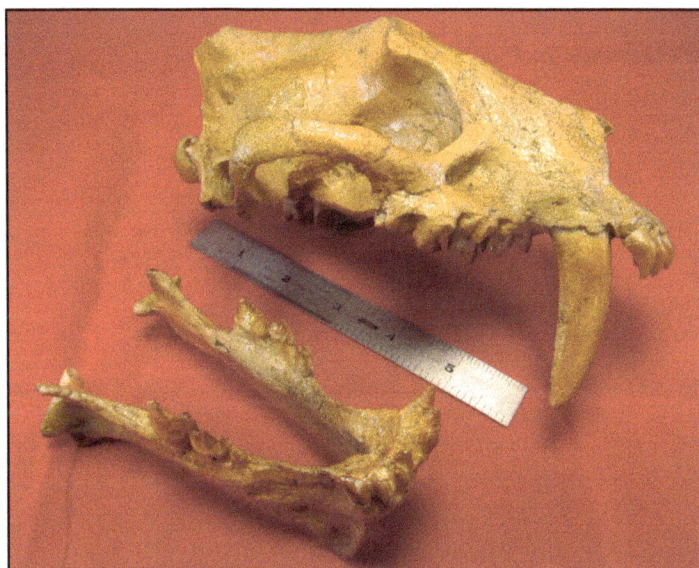

Figure 196: Fossil replica mandible and skull of *Hoplophoneus* a cat found in the sediments of the White River formation of South Dakota [OR14].

---

*   See Mitchell, J.D., *Discovering the Animals of Ancient Oregon*, p. 218.

**   *Ibid*, p. 224.

Description: The *Hoplophoneus* mandible is 5" long × 2 ¼" high × 3" wide maximum. The skull is 7" long × 4 ¼" high × 4 ½" wide. The longest sabre-tooth canine extends 2 ⅝" from the skull.

Interpretation: Relevant "18 Facts" > 1, 8, 9, 10, 13, 17. There is minimal variation in fossil cats excepting for the size of the canine teeth. It makes sense that these teeth could have varied as allowed by the inherent DNA over the 1,600 year period from the fall to the Flood. God probably chose a pair of representative cats to go on the ark from among the variety that had adapted to the ecological conditions up to that time. From those two cats developed, over a period of 4,500 years since the Flood, the variety of cats that we see today in the world.

Hesperocyon Dog

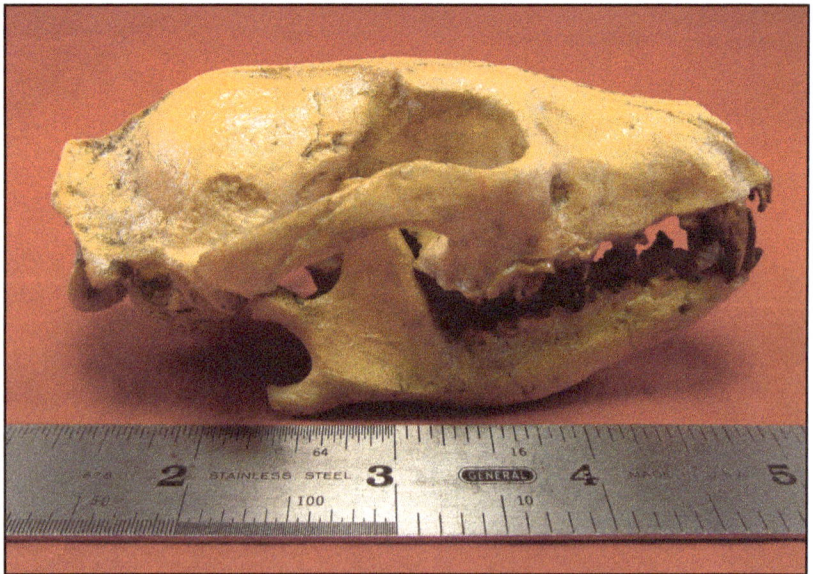

Figure 197: Fossil replica skull of *Hesperocyon* dog found in the Green River formation of Wyoming [OR13].

Description: The skull is 4" long × 2" wide × 1 ¾" high. The canine teeth extend ⅜" from the skull. This cast is of a nearly perfect fossil *Hesperocyon* skull.

Interpretation: Relevant "18 Facts" > 8, 9, 10, 14, 17. Probably the most amazing fact about the study of fossil dogs is that there is more variation in the skulls of today's artificially selected dogs than in fossil dogs. In artificial selection (breeding) humans have much more control over which animals mate than would be expected from natural conditions. The dog fossils from the Flood indicate those dogs were very similar to those extant today.

Another factor in this variation may be that there was only about one-third the time available for variation to occur in dogs between the fall and the Flood compared to the time since the Flood. Of course, the conditions prior to the Flood were much different from those after, so time may not be a significant factor in this variation.

Hyracodon Mandible

Figure 198: Partial fossil mandible of a *Hyracodon* "Running Rhino" from the White River formation, Pennington County, South Dakota [MF11].

Description: The left portion of the mandible of the *Hyracodon* has four molars intact. The right side has 5 ½ molars intact with a tooth-row length of 4 ½" and the mandible width is 2 ⅞". All of the teeth chewing surfaces show considerable wear. The bone is completely permineralized and all front teeth along with the skull are missing.

Interpretation: Relevant "18 Facts" > 1, 2, 3, 15. *Hyracodon* fossils represent an extinct rhinoceros type of animal found in America in the White River formation. The specimen above is a relatively inexpensive fossil bought as a matrix "glob" by the author to practice preparation techniques. This fossil was cracked and smashed by overburden and/or burial and transportation forces. No predator actions are indicated. As with many fossils, using biblical presuppositions, it can be assumed that the *Hyracodon* mammal's skeleton was disarticulated, transported and buried by the catastrophic actions of the Genesis Flood. Eventually, it was uncovered by the extreme erosional actions of the badlands.

Armored Mammal *Glyptodon*

Figure 199: Fossilized *Glyptodon* mammal armor scute from Argentina, South America [MF13].

Description: The *Glyptodon* scute is 1 ⅝" wide × 1 ⅞" long × ¹⁵⁄₁₆" thick. A portion of the upper left is missing. (The scutes are usually five-sided.)

Figure 200: Reconstruction of the *Glyptodon* skeleton and its armor [Houston Museum of Natural Science].

Interpretation: Relevant "18 Facts" > 8, 9, 10, 12, 13, 18. Turtles, tortoises, and crocodilians all have osteoderms or scutes that cover the exterior surfaces of their bodies to protect their inside organs. The giant crocodile *Sarcosuchus* from the rock record had huge scutes up to twelve inches long. The Armor-Backed and the Club-Tailed dinosaurs also had body armor.* The *Glyptodon* is an extinct armored mammal quite different from each of those armored animals.

Evolutionists assume that all these different kinds of animals must have evolved their armor along separate paths. The biblical creationist assumption is that God designed each animal to have armor as required to best survive. As with other defining

---

* See Mitchell, J.D., *Guidebook to North American Dinosaurs According to Created Kinds*, CEC Publications, 2014, p. 82, 95.

obviously-designed characteristics, there is much variation within each of the created kinds, but there was no evolution from an imagined common ancestor over millions of years. For these similar characteristics in different animals, a Common Designer was involved in each of their designs.

Whale Ear Bone

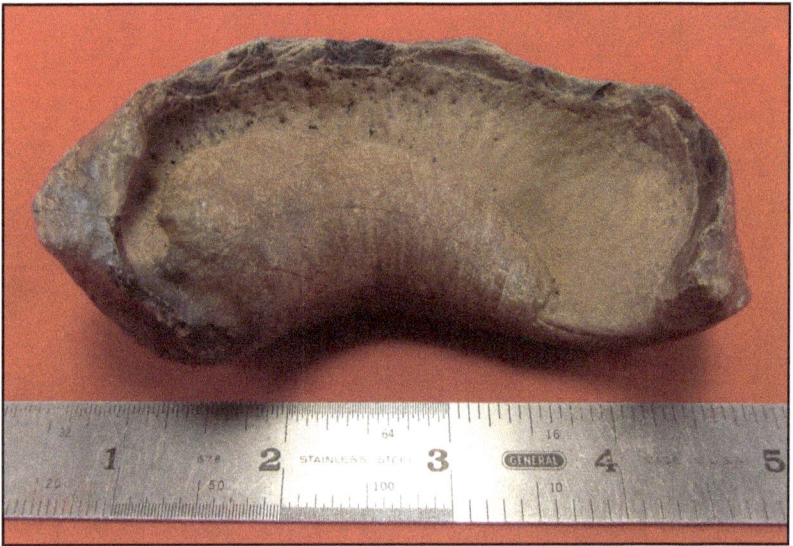

Figure 201: Fossilized tympanic bulla (ear bone) from a whale found along with Megalodon teeth in South Carolina [MARF53].

Description: The ear bone is 4" long × 1 ⅝" wide × 1 ¼" thick and is completely permineralized. The condition of the fossil is very good with fine details observable.

Interpretation: Relevant "18 Facts" > 1, 2, 3, 14, 17. According to the Bible, animals in the oceans did not come onto the ark with Noah and many survived the Flood so as to exist today. Since whales and sharks such as Megalodon are both animals of the ocean, it makes sense that their fossils would often be found together in the rock record.

The design of fossilized whale ear bones and existing whale ear bones are the same or very similar. The ear bones of whales are not at all like the ear bones of the land animals that secularists believe whales evolved from. Whales and other sea creatures were created on Day Five, a day before the land animals were created on Day Six. According to the Bible, the sea creatures all preceded the land creatures; a fact that evolutionists reject.

Megalodon Shark Tooth

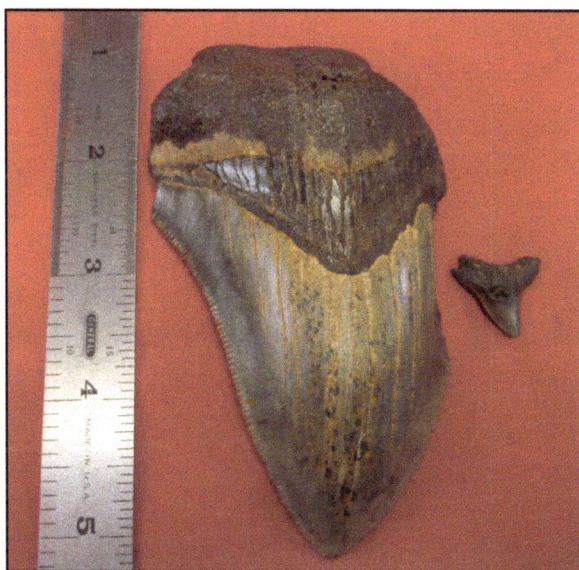

Figure 202: Giant fossil *Carcharocles megalodon* [MARF51] tooth with *Carcharhinus* (Bull shark) fossil tooth.

Description: The giant tooth is 4 ¹⁄₁₆" long × 2 ³⁄₈" wide × ¾" thick (a portion is broken away). It has 26 serrations per inch along both edges with the serrations about ¹⁄₁₆" on center. The root is about thirty per cent of the tooth length and the tooth is nearly symmetrical in shape. The smaller tooth is ¾" long × ¾" wide × ³⁄₁₆" thick with tiny serrations on both edges. The tooth crown is inclined away from the tooth centerline on the smaller specimen.

Interpretation: Relevant "18 Facts" > 8, 9, 10, 14, 17, 18. Sharks are characterized by a skeleton made of cartilage that seldom fossilizes instead of bone, so their teeth are the most common fossil found. And, the teeth, fossil and non-fossil, are found by the hundreds of thousands all over the world in sediments and on the ocean floor. It is thought that a single shark may go through 20,000 teeth in its lifetime because sharks don't chew but chomp, and they lose some teeth whenever they feed.

Carcharocles megalodon commonly known as "Megalodon," is currently thought to be extinct. Fossil remains indicate that it may have grown to be fifty feet long and twenty tons or more, much larger than the fearsome modern-day great white shark. Megalodon teeth seven inches or longer have been recovered. Complete fossil teeth four inches long and larger are commonly sold for hundreds of dollars each. Megalodon may be the same created kind as the great white shark.

Figure 203: Megalodon teeth-filled jaws in natural history museum [Credit: public domain Wikipedia].

CHAPTER FOUR

Plant Fossils

Ginkgo Leaves & Wood

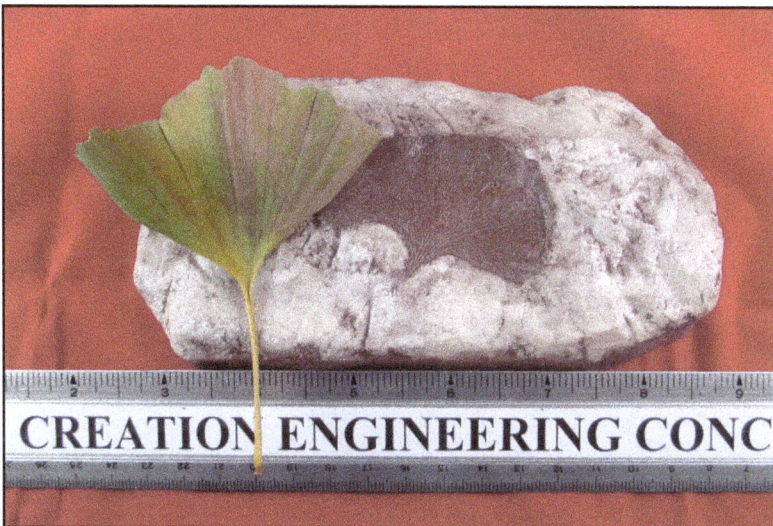

Figure 204: *Ginkgo biloba* fossil leaf replica
from an unknown location [OR22].

Description: The replica fossil *Ginkgo* leaf is 2 ½" wide × 1 ¾" high in a 6 ½" long × 3 ½" high matrix. Modern *Ginkgo biloba* leaf shown is nearly identical to the fossil.

Interpretation: Relevant "18 Facts" > 1, 6, 7, 14, 15, 17. The secular view is that the fossil leaf is 270 million years old. Since the same leaf exists today, *Ginkgo biloba* is a living fossil. The author has a live *Ginkgo biloba* tree in his backyard from which the modern leaf in the figure was obtained. The biblical creationist explanation is that the fossil leaf was rapidly buried in the catastrophic worldwide Flood at the time of Noah about 4,500 years ago.

Figure 205: Fossil leaf *Ginkgo adiantoides* from the Sentinel Butte Member, Fort Union formation, Morton County, North Dakota [OF19].

Description: The *Ginkgo adiantoides* fossil leaf is 2 ⅞" high × 1 ¾" wide and the matrix is 4 ½" high × 3" wide × ¼" thick. A portion of the leaf is broken off along with the matrix on the right hand side.

Interpretation: Relevant "18 Facts" > 1, 8, 9, 10, 17. Leaves on the author's *Ginkgo biloba* tree can be split or not split, which is an indication that this characteristic is not a defining one for determining species. *Ginkgo adiantoides* and *G. digitata* are two species names assigned to fossil leaves from the rock record along with *G. biloba*, a still living species. This variation within a kind indicates that God probably originally created one family of *Ginkgo* in the beginning. Most biblical creation scientists believe the created kind is closest to the family level, a larger variation-encompassing category than genus and species.

Figure 206: *Ginkgo* petrified (fossilized) wood sectioned and polished from Ginkgo Petrified Forest State Park, Vantage, Washington [OF9].

Description: The piece of *Ginkgo* petrified wood is 2 ¾" wide × 2" high × ¹³⁄₁₆" thick. The grain of the wood runs parallel to the bottom and top edges of the specimen.

Interpretation: Relevant "18 Facts" > 1, 2, 3, 4, 15, 17. Evolutionists teach that petrified wood is formed slowly over millions of years during which time minerals gradually replace the wood cells as the wood rots away. But, in the cases where the wood grain has been replaced with minerals to leave the same structure as before, it makes more sense that this was a relatively quick process assisted by heat, pressure and lots of water filled with minerals. The discussion of the rapid wood petrification process is found below in the section on rainbow wood from Arizona.

Petrified Wood

Figure 207: A partial section of a petrified (fossilized) tree trunk that was about one foot in diameter. Species and source location is unknown [OF8].

Description: Some of the exterior bark can be seen along the left edge and several growth rings are visible circling the center toward the upper right. The cut section has been polished and painted with a preservative. The block is 5 ½" wide × 4 ¼" high × 2" thick and weighs just over three pounds.

Interpretation: Relevant "18 Facts" > 1, 2, 3, 4, 15, 17. Petrified wood is found all over the earth. That is verification that the Genesis Flood was not a local event.

Figure 208: Petrified *Araucarioxylon* wood from the Chinle formation in northeastern Arizona, near the Petrified Forest National Park [OF14].

Description: The specimen is 3 ¾" wide × 5 ½" long × 2 ½" thick and weighs almost two pounds. The surface looks like wood but has been completely turned to stone.

Interpretation: Relevant "18 Facts" > 1, 2, 3, 4, 15, 17. The specimen is an example of "Rainbow Wood" and is from a tree with the genus name *Araucarioxylon*. The biblical creationist notices that the petrified trees of the large area encompassing the Petrified Forest National Park are all laying down flat and with many largely in a preferred orientation. Also, the tree branches are stripped and many of the logs are broken into short chunks. All of the remnants have been made into petrified wood by molecular replacement of the original wood with the mineral silica.

The rainbow wood forest is best explained as having been transported to where it currently rests by a large amount of water or mud. It is, as seen today, a result of the global Flood cataclysm at the time of Noah. The fossilization process can also best be attributed over such a large area to special conditions only in place during this one-time year-long Flood.

Scientific experimentation has shown that rapid petrification of wood takes place in five stages:

1. Entry of silica in solution or as a colloid into the wood.
2. Penetration of silica into the cell walls of the wood's structure.
3. Progressive dissolving of the cell walls which are at the same time replaced by silica so that the wood's dimensional stability is maintained.
4. Silica depositions within the voids within the cellular wall framework structure.
5. Final hardening (lithification) by drying out.*

Figure 209: Petrified *Araucarioxylon* wood tree chunks at the Petrified Forest National Park in Arizona [Credit: National Park Service].

---

* Snelling, Andrew A, *Creation* 17, no 4, September 1, 1995.

Other Fossil Leaves

Figure 210: Fossil leaf imprint from the John Day
formation in Wheeler County, Oregon [OF3].

Description: Incomplete fossil leaf is about 1 ¾" wide. Matrix is
2 ⅞" wide × 1 ¾" high × ⅝" thick. On the reverse side of this
specimen is an imprint of a willow leaf ½" wide.

Interpretation: Relevant "18 Facts" > 1, 2, 3, 17. In the John Day
Fossil Beds of Central Oregon large masses of plant fossils (mostly
leaves) are found. Among the common leaves found are birch, elm,
maple, oak, hickory, poplar, alder and beech. The leaf in the figure
above looks to be from either a birch or poplar tree.

The leaves in these plant fossil masses are very similar, or identical, to the leaves we find on the same types of trees today. The
author's biblical creationist explanation for these leaf masses is that
they were formed rapidly by massive movements of water-laden
sediments during the worldwide Flood about 4,500 years ago.

Figure 211: Fossil *Salix* (willow) leaves from Eastern Oregon [OF4].

Description: Largest *Salix* leaf is 1 ⅞" long × ⁷⁄₁₆" wide. The matrix is 1 ¾" × 2 ½" × ¼" thick.

Interpretation: This leaf fossil is also possibly from the John Day Fossil Beds of Oregon. To form it had to be rapidly and completely buried in some way generally not seen today. The likely event was the one-year-long global Flood described in the Bible that happened about 4,500 years ago.

Figure 212: Fossil leaf in mudstone from the Owyhee Mountains in Idaho. This is likely a *Salix* (willow) leaf) [OF20].

Description: The *Salix* leaf is 5 ⅝" long × 1 ⅜" wide and the matrix is 5" wide × 6 ¾" high × ⅞" thick. The fossil was broken and glued back together at a joint about two inches from the stem end. The leaf has been coated with shellac for preservation. There are nine "teeth" along the left edge and the right edge is 75 percent missing.

Interpretation: Relevant "18 Facts" > 1, 15, 17. Willows are a flowering plant that the Bible says God created on Day Three of the creation (See Genesis 1:11-12). Fossil leaves of many kinds of plants are commonly found in the clays, volcanic ash and other fine sediments that resulted from the global Flood at the time of Noah. These fossil willow leaves are the same as today's leaves, yet there is little evidence that leaves are still being fossilized in lakes or rivers today.

Figure 213: Fossil *Salix* leaf from a location in the state of Utah [OF21].

Description: The leaf is 2 ½" long × ⅜" wide and the matrix is 1 ½" wide × 3 ¾" long × ⅜" thick. Under a magnifying glass, teeth are visible along the edges of the leaf.

Interpretation: Relevant "18 Facts" > 1, 15, 17. There is no discernable difference between this leaf from the rock record and extant *Salix* leaves growing on willow trees today. This is another example of a living fossil!

Figure 214: Fossil leaf *Zelkova* (elm) in limestone from Erdobenye, Zemplin Mountains, Hungary [OF22].

Description: The *Zelkova* (elm) leaf is $^{13}/_{16}$" long × $^{9}/_{16}$" wide and the matrix is 3" long × 2 ¼" wide × $^{11}/_{16}$" thick. There is also a $^{9}/_{16}$" long piece of unknown organic matter located at the bottom of the piece.

Interpretation: Relevant "18 Facts" > 1, 8, 15, 17. This delicate fossil elm leaf has the same basic shape and design as leaves on elm trees of today, yet is thought to be 25 million years old by evolutionists. In reality, we see variation within plant kinds, but no macroevolutionary march from one plant kind to another entirely different kind. That reality is the same in the extant plants of today as well as in fossil plants from the rock record.

If *Zelkova* fossil leaves are formed into fossils in a continuing uniformitarian manner over millions of years, then we should see them in consolidated sediments that we know are 100, 500 or 1,000 years old. But, fossilization does not seem to happen according to the gradual conditions of evolutionary theories laid out in secular textbooks. It makes sense that these leaves were fossilized in a non-uniformitarian event. The cataclysmic Genesis Flood is the best explanatory candidate!

Figure 215: Fossil leaf *Acer* (maple) in lignite from the Brown Coal Mine in Bilina, Bohemia, Europe [OF28].

Description: The *Acer* leaf is 1 ¾" wide × 1 ¹¹⁄₁₆" high. The matrix is 2 ½" wide × 2 ⅝" high × ⁷⁄₁₆" thick. The stem extends from the bottom of the leaf ⁵⁄₁₆" and is ¹⁄₃₂" wide. A portion of the center-right of the leaf is missing from the fossil.

Interpretation: Relevant "18 Facts" > 1, 8, 15, 17. This fossil leaf species *Acer tricuspidatum* is basically identical to the extant maple leaf species *Acer rubrum* that today ranges from the maritime provinces of Canada to Florida and west to Texas and Minnesota. This is another of the hundreds of identified living fossils.

Figure 216: Fossil plate from the John Day formation of the John Day Fossil Beds of Oregon, with two *Metasequoia* leaves [OF5].

Description: The fossilized *Metasequoia* leaves are $^{11}/_{16}$" wide maximum with one 2 ½" long and the other 1 ½" long. The matrix is 3" × 3 ½" × $^5/_{16}$" thick. Each needle is arranged opposite another corresponding needle on the other side of the stem.

Interpretation: Relevant "18 Facts" > 1, 2, 3, 17. The tree represented by the twigs in the figure above was only known from fossils until it was found growing in China in the 1940s. Also known as dawn redwood, it is one of the more famous examples of a living fossil with some evolutionists exclaiming that its discovery was "like finding a live dinosaur."

Metasequoia trees are now cultivated all around the world. The giant *Sequoia* trees found in California have their needles placed on the stems in an alternate fashion rather than opposite like the *Metasequoia*. The two types of trees also have slightly different seed cones as will be described in the next section.

Sequoia Seed Cone

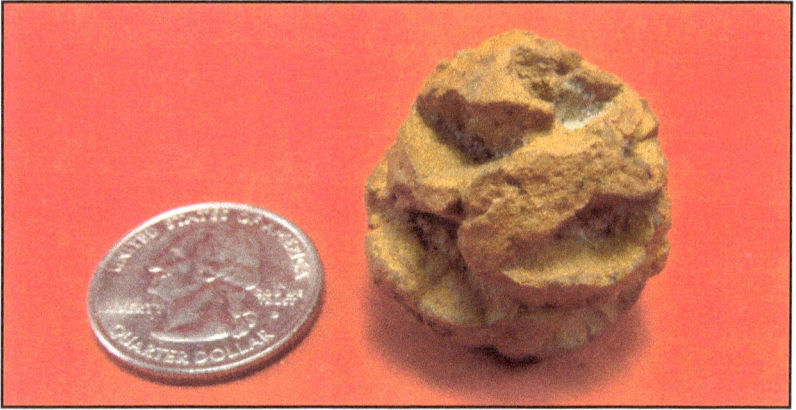

Figure 217: Fossilized *Sequoia* seed cone from the Hell Creek formation of South Dakota [OF27].

Description: The permineralized *Sequoia* seed cone approximates a 1 ⅛" diameter sphere. The scales are staggered instead of vertically arranged. The typical scale is diamond shaped and is ⅝" wide × ¼" high.

Interpretation: Relevant "18 Facts" > 1, 2, 3, 8, 12, 14, 17. Both *Sequoia* and *Metasequoia* trees are living fossils that are similar but different in a number of important ways. The seed cone design and the stem/needle design are two of the differences. When the seed cones for these trees are found as fossils, the seeds have usually already fallen out of the cones. This fact provides some insight for the creation scientist for what time during the year-long Flood that the cones were buried.

God's creation is amazing in millions of ways. Consider that living things have what is needed for reproduction already packed into their DNA. *And God said, let the earth bring forth grass, the herb yielding seed, and the fruit tree yielding fruit after his kind, whose seed is in itself, upon the earth: and it was so* (Genesis 1:11). So in this verse God said He was providing the fruit and seeds for reproduction and for food. And, even today, plants provide us our food in abundance either directly or indirectly. Bible verses always provide unique and profound knowledge that is found nowhere in evolutionary theories.

Alethopteris Fern Frond

Figure 218: Fossil seed fern fronds from *Alethopteris* in a black shale slab from Pennsylvania [OF7].

Description: The largest *Alethopteris* frond is 5" long × 1 ½" wide. Leaflets are attached to the stalk along widened bases and are oriented obliquely to the stalk. The largest leaflets are ¼" wide × ¾" long. The shale matrix is 5 ½" × 7 ¼" × ¼" thick with frond imprints on both sides of the slab and presumably inside as well.

Interpretation: Relevant "18 Facts" > 1, 2, 3, 15, 17. Evolutionists believe flowering plants evolved from plants like these *Alethopteris* ferns. But, ferns grow in warm, swampy areas and it could be expected that they would often be buried by the Flood lower in the sediments compared to flowering plants.

Lepidodendron Bark Patterns

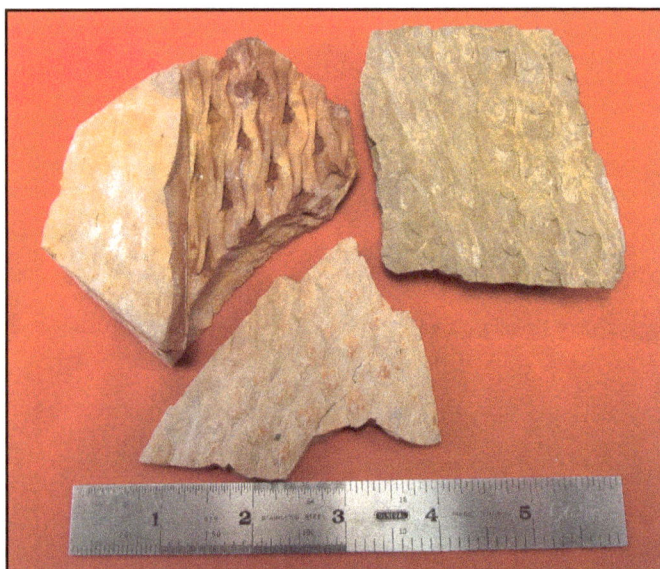

Figure 219: Three specimens of fossil *Lepidodendron* bark from lycopod plants from the Upper Silesia formation near Czerwionko, Poland [OF10-1, OF10-2, and OF10-3].

Description: Fossil *Lepidodendron* specimen OF10-1 (left) is 3 ½" long × 2 ¾" wide × 1" thick. Its triangular shaped scars are in rows ½" apart on 1 ⅛" centers and are ⅛" deep. Fossil specimen OF10-2 (right) is 3" long × 2 ¾" wide × ⅝" thick with an irregular scar pattern. Fossil specimen OF10-3 (lower) is 3 ¼" long × 2 ½" wide × ¼" thick. Its diamond shaped scars are in rows ⅜" apart on ⁷⁄₁₆" centers.

Interpretation: Relevant "18 Facts" > 1, 8, 14, 17. The *Lepidodendron* fossils are the most common lycopod plant found in the rock record, they are noted for their scale-like bark patterns, as seen in the figure. These patterns vary not only species to species, but also within the height of individual lycopod trees that have been found to have grown to be over one hundred feet tall and with trunks up to six feet in diameter.

The differences in *Lepidodendron* bark patterns are similar in amount of variety to those seen in today's conifer trees, although lycopods are more like reeds rather than trees. Today's lycopods are represented by small club mosses that never approach the size of fossil lycopods.

The biblical creationist asks: Did God create all of the variety seen in the *Lepidodendron* bark designs at the beginning, or are the variations due to adaptions, allowable by the DNA, that occurred during the 1,600 years between the creation and the Flood?

Lycopods and Petroleum

Figure 220: Three specimens of bark fossils from lycopod plants with petrochemical smell and color from the Upper Silesia formation of Poland [OF10-5, OF10-4, and OF10-6].

Description: Fossil specimen OF10-5 (left) is 3 ½" long × 2" wide × ³⁄₁₆" thick. Its scars are in rows ³⁄₃₂" apart and on ⁵⁄₁₆" centers. Specimen OF10-4 (center) is 1 ⅜" square × ¾" thick. Its diamond shaped scar pattern tends to vary as it spirals around a circumference. Specimen OF10-6 (right) is 2 ¾" long × 1 ¼" wide × ⁷⁄₁₆" thick with an irregular scar pattern.

Interpretation: Relevant "18 Facts" > 1, 8, 12, 14. The secular view is that fossil lycopod plants are indicators of a particular time period millions of years ago. Many of the fossil lycopods are commonly found associated with coal-bearing strata. Coal miners often mistakenly believe lycopod bark fossils are snake fossils because of the scale-like patterns found on them.

Creationists propose that the petroleum products extracted from the earth's crust today are the consequence of massive plant matter mats that were modified to their present form by temperature and pressure after rapid burial during the global Flood. Plant fossils as well as other types of fossils are often found in coal seams, and that fact meshes with the creationist hypotheses better than the slow accumulation of vegetable matter in peat bogs, the best evolutionary story for coal, gas, and oil.

Neuropteris Fern

Description: The *Neuropteris* leaflet is ¹⁵⁄₁₆" wide × 2 ¾" long and the nodule is 2" wide × 3 ¹⁄₁₆" long × ⅞" thick. A ⅛" wide curved tear in one side of the leaflet is visible in both halves of the fossil.

Figure 221: Fossil seed fern leaflet *Neuropteris* in an ironstone nodule from the Mazon Creek fossil bed in Grundy County, Illinois [OF12].

Interpretation: Relevant "18 Facts" > 1, 2, 3, 14, 15. The Mazon Creek fossil beds are of interest because both hard and soft tissues of many different kinds of animals and plants are preserved in concretions there. The formation consists of about one hundred feet of shale and most of the fossils are found in its lower fifteen feet or so. The nodules are called "ironstone" because dissolved iron from the water was involved in the fossilization process, and the iron is noticeable in the concretions. The *Neuropteris* is the foliage of an extinct type of fern that reproduced by seeds instead of spores.

The formation of these Mazon Creek nodules required rapid burial and unusual conditions seldom, if ever, experienced today. If the process of saving the fossil was a gradual one the item inside the nodule would have deteriorated to nothing. The seed fern that this *Neuropteris* leaflet represents grew in abundance prior to the worldwide Flood, but was not able to re-establish itself after the Flood.

Pecopteris Fern

Figure 222: Common Tree Fern frond *Pecopteris* from the Mazon Creek fossil bed in Grundy County, Illinois [OF16].

Description: The *Pecopteris* frond is 1 ³/₁₆" long × ³/₈" wide. The larger nodule piece is 1 ¾" wide × 2 ¾" long × ¹¹/₁₆" thick. The smaller iron stone piece is 1 ³/₈" diameter × ¼" thick. The shape of this fossil-bearing nodule is highly unusual.

Interpretation: Relevant "18 Facts" > 1, 2, 3, 14, 15. A Tree Fern plant left this fossil frond segment as testimony to the Genesis Flood. *Pecopteris* is considered extinct today as it was unable to re-establish after the Flood.

Annularia Leaf

Figure 223: Fossil *Annularia* leaf whorl from a *Calamites* tree from the Mazon Creek fossil bed in Grundy County, Illinois [OF15].

Description: The *Annularia* leaf whorl imprint is 1 ⅞" wide × 1 ¼" high and the nodule half-section is 2" diameter × ⁷/₁₆" thick. The largest leaf in the whorl is ⅛" wide × 1 ⅛" long.

Interpretation: Relevant "18 Facts" > 1, 3, 14. The *Annularia* leaf has a separate genus name from the *Calamites* tree on which it grew.

This indicates how difficult it can be to properly analyze the fossils that are the result of the worldwide cataclysm. Fossils of the now extinct *Calamites* indicate that it grew to heights of as much as one hundred feet. The much smaller horsetail rush (*Equisetum*) is a similar plant to *Calamites* that still exists today.

Calamites Tree Stem

Figure 224: Fossil of a natural cast of a flattened *Calamites* tree stem from an Alabama coal mine [OF17].

Description: The fossil cast (not a replica) is 2 ¾" wide × 4 ⅛" long. It is 1 ⅞" thick and so has been compressed ⅞" from its original circular cross-section. The *Calamites* distinct rib like ring is ⅛" wide and the vertical ribs that run longitudinally are ¹⁄₁₆" wide. The fossil is carbonized on the visible end in the figure.

Interpretation: Relevant "18 Facts" > 1, 3, 14. This fossil not only provides a record of rapid burial of a tree stem, but also shows that it was buried under enough sediment to flatten the original round cross-section to an elliptical shape. Some complete fossil trees show the same type of elliptical cross-sections for the same reasons as seen in the figure below. Many evidences of the close connection between plant matter and coal have been found in coal mines.

Figure 225: Large fossilized tree on display at the Creation Evidence Museum in Glen Rose, Texas that was flattened by Flood sediments prior to petrification.

Fossil Flower

Description: The flower petals in Figure 226 are in the shape of a five-pointed star and the flower is $5/32$" × $5/32$". The matrix is 1 ¼" wide × 1 ¾" long × $5/16$" thick.

Interpretation: Relevant "18 Facts" > 8, 9, 10, 11, 13, 17, 18. According to the Bible, God created plants on Day Three ahead of flying and sea creatures on Day Five, as well as animals and humans on Day Six. According to evolutionary theories, flowering plants did not evolve until long after land animals appeared. This

Figure 226: Fossil flower on matrix from the Green River formation of Rio Blanco County, Colorado [OF26].

is just one of the many foundational presuppositions of secularism that is counter to God's plainly written Word.

The rock record reveals plant and flower variety but nowhere shows evidence of macroevolution. According to the book of Romans chapter one, creation is obvious to everyone in what is seen. If evolution were true, it should be obvious, but it can be found nowhere except in the minds of Creator-rejecting men.

CHAPTER FIVE
Trace Fossils

Coprolite Trace Fossils

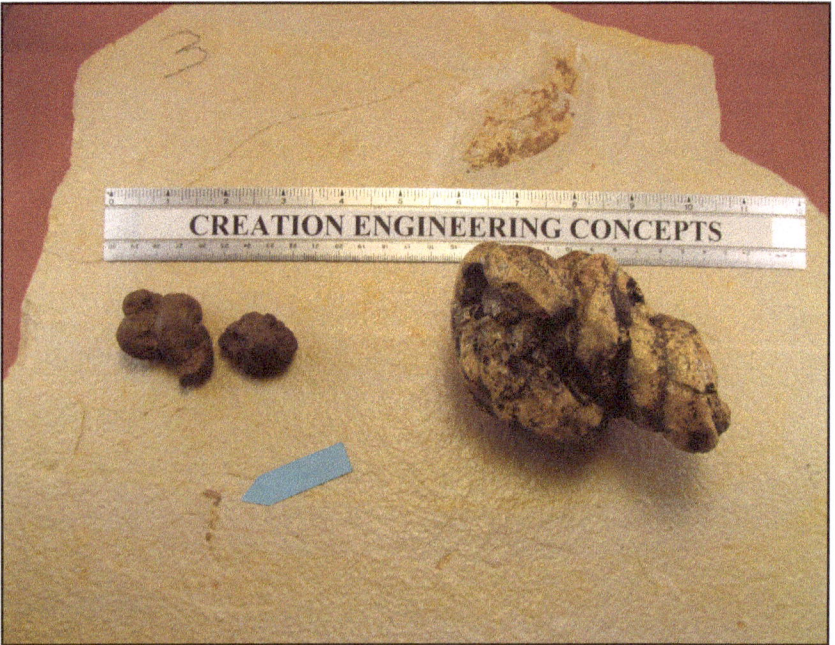

Figure 227: A grouping of trace fossils: Animal Coprolite (2 pieces), fish coprolite in limestone matrix from Wyoming (blue arrow), replica dinosaur coprolite from North America [DF6, DR11].

Description: The larger specimen of coprolite from the unknown animal (perhaps a dinosaur) is 1 ¼" wide × ¾" thick × 1 ¾" long. It has a relatively smooth surface. The smaller coprolite piece is about 1" diameter × ¾" high. It has a number of small protuberances that could be interpreted as fossilized plant seeds.

The dinosaur coprolite replica is 3" wide × 2 ½" thick × 4" long. All of the animal coprolite specimens resemble feces deposits from living animals. The fish coprolite in limestone matrix from the Green River formation near Kemmerer, Wyoming is ³⁄₃₂" diameter × 1" overall length. It is located in the same plane as the fossilized fish at the top of the matrix.

Interpretation: Relevant "18 Facts" > 1, 15. The fossilized remains of feces are called coprolites. Generally speaking coprolites are not found close by to skeletal animal fossils. However, one exception is that fish coprolite is very common alongside fish fossils in the limestone layers of the Green River formation. In the case of dinosaur coprolites there is no agreement among experts about how to identify them. Some coprolites originally identified as being from dinosaurs were later attributed to crocodiles and other animals.

Ignoring the problems in correlating coprolites to their sources, it is generally agreed that only rapid burial under water-laden sediments can explain their existence. These are the conditions we can expect were in place during the Flood that is described in the book of Genesis.

Gastrolith Trace Fossil

Description: The gastrolith stone in Figure 228 is 2" wide × 2 ¼" long × 1 ½" high with an irregular shape and a smooth surface.

Interpretation: Relevant "18 Facts" > 2. Gastroliths (Greek: Stomach stone) are rocks and pebbles ingested into the stomach, or elsewhere in the intestinal tract, to help some animals with digestion. They are found in living animals today such as crocodiles, alligators and some birds. Since the early 1960s paleontologists have believed that certain

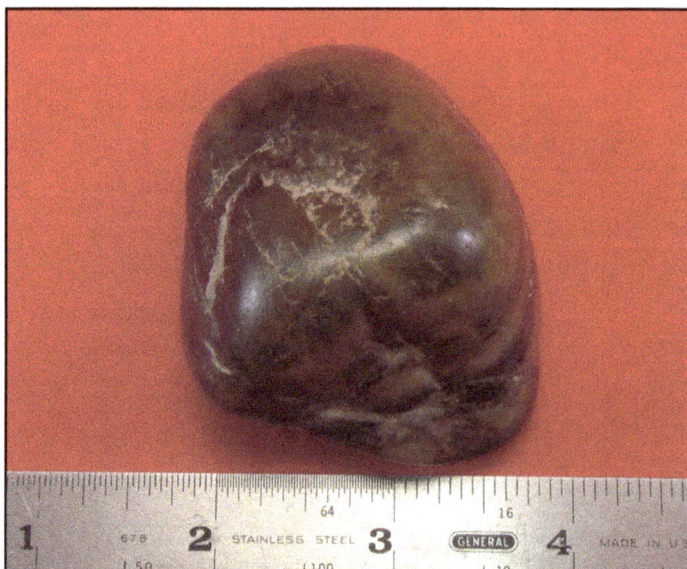

Figure 228: Dinosaur gastrolith fossil from the Morrison formation of North America [DF13].

groupings of stones found within or near fossilized dinosaur remains were used by the dinosaurs in this manner.

There is little doubt that some dinosaurs used gastroliths to aid in digestion. What is doubtful is that all of the multitudes of polished rocks promoted as dinosaur gastroliths are authentic. Some fraudulent stones on the market today are likely to have been polished using mechanical means. The gastrolith in the figure is thought to be authentic for the following reasons:

1. It was found together with other similar rocks over 50 years ago and has been in the author's collection continuously since.

2. It is rounded and polished in the manner characteristic of authenticated museum dinosaur gastrolith specimens. Dents have been smoothened just like the more exposed surfaces.

3. It has dents and scratches characteristic of those expected from the stone coming in contact with the sharp corners of other freshly swallowed stones.

Dinosaur Footprint

Figure 229: Cast of a small three-toed dinosaur footprint from Massachusetts [DR7].

Description: The dinosaur footprint is 3" wide × 4" long × ½" maximum depth. The matrix is 5 ¼" wide × 5 ½" long.

Interpretation: Relevant "18 Facts" > 1, 12, 15, 16. Many dinosaur footprints in stone have been discovered over the years, but dinosaurs are not found standing in the footprints so it is not easy to correlate a footprint to a particular dinosaur. This footprint is obviously of a small to medium sized three-toed individual, but other than the location and rock layer where it was found, that is the extent of what can be determined with any certainty.

Creationists and evolutionists have a number of theories for how the footprints and associated trackways were formed, but why would it be easier to believe these trace fossils are one hundred million years old rather than 4,500 years old? Which belief requires more faith?

Dinosaur Footprint

Figure 230: *Magnoavipes caneeri* dinosaur footprint cast from Dinosaur Ridge in Morrison, Colorado [DR23].

Description: The *Magnoavipes* cast was taken from a footprint of the many examples found at the main track site at Dinosaur Ridge near Morrison, Colorado. The footprint is 10" wide × 9" long × ¾" maximum depth. The matrix is 11 ½" wide × 9 ³⁄₁₆" long.

Interpretation: Relevant "18 Facts" > 1, 12, 15, 16. *Magnoavipes caneeri* is the genus/species name that has been applied to the footprint believed to have been made by a relatively small Tyrant Bipedal (theropod) dinosaur.

At the main Dinosaur Ridge track site, footprints from three different types of dinosaurs and a type of crocodile have been identified. At this site at least 350 individual tracks were made by about 40 individuals, with the majority thought to have been made by a Duck-Billed kind of dinosaur.*

---

* Lockley, Martin and Marshall, Clare, *A Field Guide to the Dinosaur Ridge Area,*

The conditions that preserved these tracks are much more likely to have been from a catastrophe than from uniformitarian conditions over deep time. The millions of dinosaur footprints scattered over the surface of the earth exist today because of the special conditions of the Flood at the time of Noah!

Figure 231: Dinosaur and crocodile tracks at Dinosaur Ridge near Morrison, Colorado. The footprints have been colored by a water-diluted charcoal mix for visibility.

Trilobite Tracks

Description: The slate slab is 2 ⅜" wide × 2 ¾" high × ⅛" thick. The trilobite tracks cover the total distance of 2 ¾" or the total height of the slab. The trackway is nominally ½" wide and there are about 28 prints per side.

Interpretation: Relevant "18 Facts" > 1, 2, 3, 12, 15, 16. The author does not know, nor is it likely that anyone knows, the genus of the

---

Friends of Dinosaur Ridge, 2014.

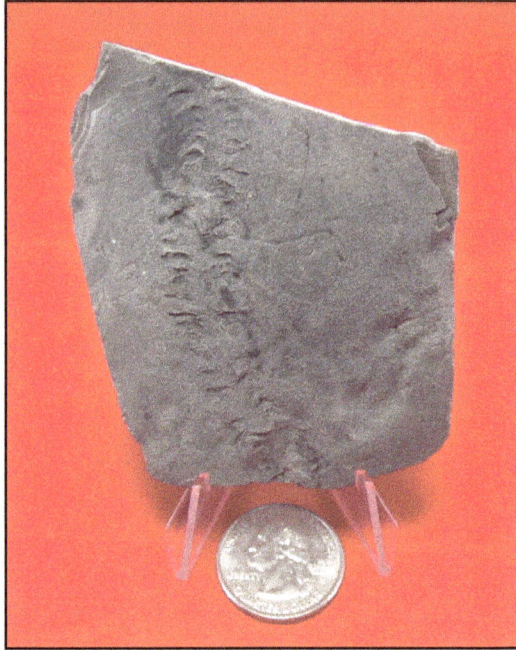

Figure 232: Trace fossil slab with trilobite tracks from the Hale formation in Washington County, Arkansas [OF18].

trilobite that made this fossil trackway since trilobites are seldom, if ever, fossilized with their tracks. If it is assumed that the trilobite was about ½" wide there would be a large number of genera that could fit the tracks. Trilobite tracks are sometimes given the genus name of *Cruziana*

Dinosaur Eggs.

Description: The larger replica dinosaur egg in Figure 233 is 5 ⅛" wide × 6 ¼" high. It is embedded in a portion of the rock matrix in which it was discovered. The smaller replica dinosaur egg is 2 ⅜" wide × 6" long × 2" high at the large end. The egg is oblong and has been crushed about ½" deep diagonally along its length.

Interpretation: Relevant "18 Facts" > 1, 12, 15. Dinosaur eggs are

Figure 233: Cast replica dinosaur eggs: Upper egg is from a Duck-Billed dinosaur from China and the lower egg is from a Tyrant Bipedal dinosaur from the Gobi Desert of Mongolia [DR16, DR6].

similar to dinosaur footprints in that it is difficult to know for sure which eggs go with which dinosaur kind. The eggs found so far are mostly nearly spherical or oblong in shape like those shown in the figure above.

Evolutionists envision some sort of slow fossilization process of these eggs. Creation scientists are largely restricted to putting forth speculations, since no one was there to witness the process, but agree that rapid burial is a more satisfactory explanation that matches the expected results of the global Flood.

Dinosaur eggs had hard shells like birds rather than the leathery shells used by many of today's reptiles. This is evidence that indicates dinosaurs were not the same created kind as reptiles.

Dinosaur Egg Shell Fragment

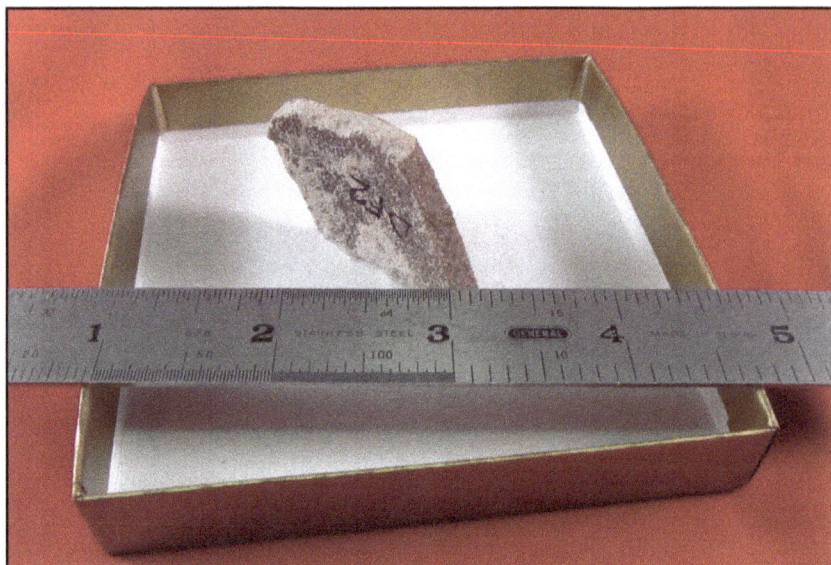

Figure 234: Fossilized dinosaur egg shell fragment [DF2].

Description: Egg shell fragment is ³⁄₁₆" thick × 1 ¼" × 1 ¼" and is part of a Long-Necked dinosaur egg from Argentina.

Interpretation: Relevant "18 Facts" > 1, 2, 3, 4, 15. Dinosaur egg shell fragments are much more common than are complete dinosaur eggs, but identifying the kind of dinosaur that an egg shell fragment came from is largely a speculative exercise. The conditions of a year-long worldwide Flood could be expected to fossilize eggs, just as they made possible millions of other invertebrate, vertebrate, plant and trace fossils.

Raindrop Trace Fossils

Description: The concave raindrops in OF29-1 range in size from ¹⁄₁₆" to ³⁄₈" diameter. There are about 12 raindrop impressions per square inch of mudstone surface. The convex raindrops in OF29-2 range in size from ¹⁄₁₆" to ¼" diameter. The matrix size for

Figure 235: Concave rain drop fossils (top) and convex rain drop fossils (bottom) in mudstone from the El Pueblo track site in the Sangre de Christo formation of northeastern New Mexico [OF29-1, OF29-2].

OF29-1 is 9" wide × 6 ¾" high × ¾" thick. Matrix OF29-2 is 7" wide × 4 ⅝" high × ⁵⁄₁₆" thick.

Interpretation: Relevant "18 Facts" > 1, 12, 15, 16. These trace fossils are from a location where numerous animal tracks have also been discovered. The secular explanation for the animal footprints, and these raindrop trace fossil impressions in the New Mexico rock record is, "a major terrestrial transference flood event." Another secular answer for these fossils is, "repetitious and intensive rainfall episodes." According to the evolutionary paradigm, these events took place hundreds of millions of years ago.

The biblical creationist interpretation is that these concave and convex raindrops were made during the Genesis Flood just 4,500 years ago. Scientific experiments witnessed by the author prove that mud can turn into rock relatively quickly and does not take millions of years. So, if uniformitarianism is valid, why don't raindrops regularly freeze into rock today? Why can't we find in the rock evidences of every monsoon over the past fifty years? The author believes it is because these fossils were made only under the very unusual circumstances of the global Flood. Secularists continue to accept more and more catastrophe into their explanations for the causes of what is found in the earth's crust, but their atheistic religion that rejects the accuracy of the Bible blocks them from considering a global Flood.

Water Ripple Trace Fossil

Figure 236: Water ripples trace fossil in mudstone from the El Pueblo site in the Sangre de Cristo formation of northeastern New Mexico [OF29-3].

Description: The mudstone specimen shows traces of water-caused multi-directional waves. One wave trace is oriented at about 15 degrees off horizontal and the other wave trace is close to vertical. The distance between ridges is 2" center-to-center for the

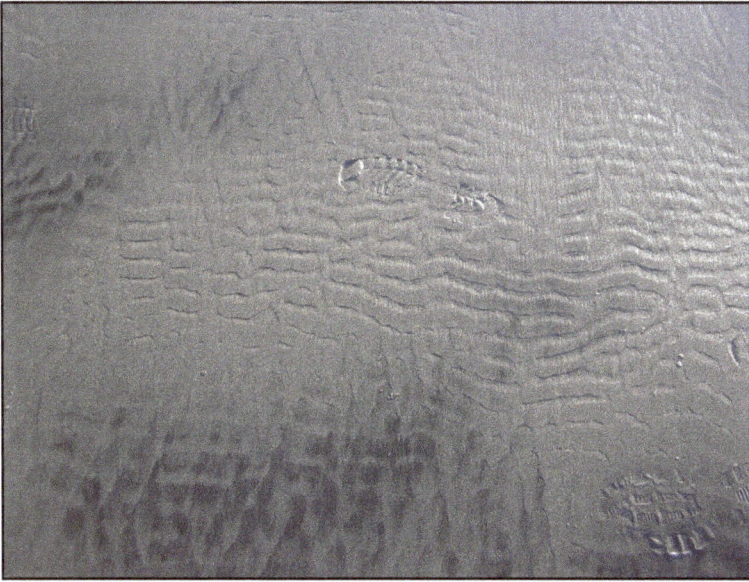

Figure 237: Multi-directional ripple marks in wet beach sand along the Oregon coast (note the man's shoe prints for scale). Today, these are never permanent and will be gone with the next wave or high tide.

Figure 238: Fossil ripple marks in sandstone at Dinosaur Ridge near Denver, Colorado.

horizontal wave trace, and 1 ½" center-to-center for the vertical wave trace. The height of the traces is about ³⁄₁₆." The rock piece is 6 ½" wide × 7 ¼" high × ⅝" thick.

Interpretation: Relevant "18 Facts" > 1, 12, 15, 16. The movement of mineral particles in a current creates ripple marks on the sediment surfaces over which they travel. Ripple marks form in various shapes, sizes and patterns that result from the type of water or wind action that acts on them. A good laboratory for studying the formation of ripples is an ocean beach.

Secularists are familiar with the many variables that cause all the different ripple designs, but they do not have good explanations for why ripples are frozen in the rock record. The biblical creationist believes they are there because of the unusual conditions and actions of the Flood at the time of Noah that are described in the Bible.

Fossil Clam Borings

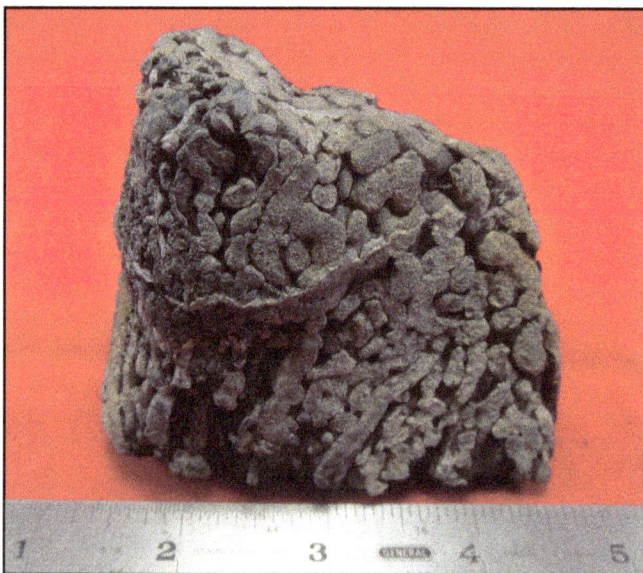

Figure 239: Fossil clam boring traces in wood (*Toredo* wood) from the Empire formation in Coos County, Oregon [OF23].

Description: The block of fossil *Toredo* clam borings is 3 ¼" wide ×
2 ⅝" high × 1 ¼" thick. The fossil borings range from ¹⁄₁₆" wide to
¼" wide and the longest borings are ¾" long. There is very little left
of the wood in which the clams bored.

Interpretation: Relevant "18 Facts" > 1, 2, 3, 17. These borings were
made by a bivalve called the *Toredo* clam. The *Toredo* is called a "ship-
worm" by mariners because of the havoc they wreak, even today,
on wooden boats and docks. The borings made in the fossil are preserved by minerals in the sediment that in-filled the bored holes.

Because the clams that made the fossilized borings have not
been preserved, this is a trace fossil. There is little if any difference
between the *Toredo* of today and the ones that made the fossil borings
in the figure above prior to the Genesis Flood about 4,500 years ago.

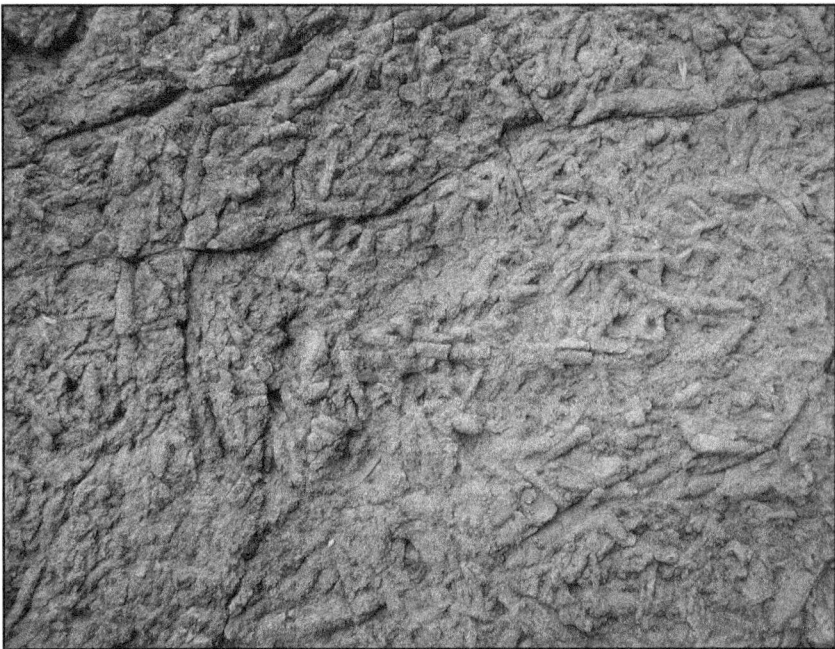

Figure 240: Trace fossil worm boring castings along
the Colorado River in the Grand Canyon.

Worm Burrow Cast

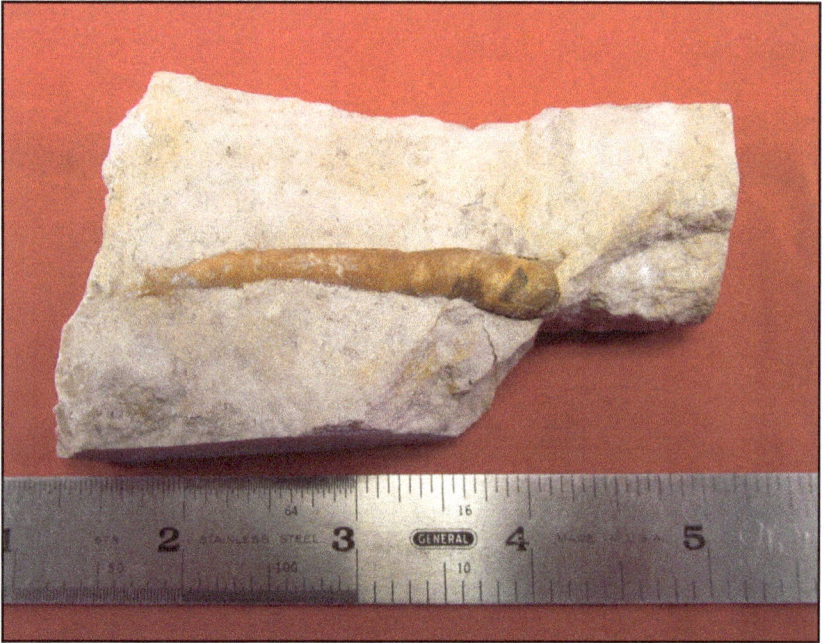

Figure 241: Worm burrow cast in limestone from the Stewartville formation near Greenleafton, Minnesota [OF24].

Description: The trace fossil is ⁵⁄₁₆" maximum diameter × 2 ⅛" long. The limestone matrix is 2" wide × 3 ½" long × 1 ¹⁄₁₆" thick. The cast has been naturally stained with iron minerals.

Interpretation: Relevant "18 Facts" > 1, 5, 12, 15, 16, 17. The processes for strata deposition and fossil burial and preservation that provided this worm burrow cast were different from those in place today. Worm burrows are seldom, if ever, fossilized by the processes we observe in the present. The principle of uniformitarianism ("the present is the key to the past") that is foundational to secular geology and paleontology is not supported by observational science.

Worm Burrow Cast

Figure 242: Worm burrow cast from the Decorah Shale near Rochester, Minnesota [OR25].

Description: The trace fossil is ⁹⁄₁₆" diameter × 5 ⁹⁄₁₆" long. The cast was broken in two and has been glued back together.

Interpretation: Relevant "18 facts" > 1, 5, 12, 15, 16, 17. The extent of worm burrow fossils in sediments indicates a process whereby the worms were able to make their normal compliment of burrows in relatively soft materials. The environment where they were living was rapidly and completely buried, the animals were destroyed, and the spaces were infilled with other materials all before erosion and other processes could destroy any trace of them. The fossils described and interpreted in this book all required large amounts of wet sediments and very rapid burial. What could have provided these conditions? The global Flood is the only answer.

Glossary

Note: Words that are **bold** within a term definition have their own glossary entry.

Abdomen: Rear or lower part of the body in many animals.

Amber: The hardened fossilized resin or exudate from certain plants, mainly coniferous trees.

Anterior: Toward the front end of an animal.

Aperture: The opening surrounded by the shell margin in mollusks.

Aptychi: In fossils, platelike structures that were possibly the jaws of mollusks such as ammonites.

Aragonite: A crystalline form of calcium carbonate.

Articulated: Fossil bones found in the same orientation as they were when the animal was alive.

Axis: In trilobites, the ridge line that runs from **anterior** to **posterior**, down the midline; in vertebrates, the first or second element of the backbone.

Beak: A prominent curved structure.

Benthonic: Living on the sea floor.

Biblical creationist: One who studies God's General Revelation (the creation) while accepting the Bible as being God's accurate and inerrant Special Revelation.

Biblical paleontology: The study of **paleontology** strictly applying biblical **presuppositions**.

Binomial nomenclature: The traditional system of two names, genus and species, for identifying organisms, both living and extinct.

Biology: The **science** that studies plants and animals.

Bipedal: Using two limbs for walking, like people, birds, and some **dinosaurs**.

Bone bed: A portion of a **sedimentary rock** layer that has a large number of **fossil** bones or fragments.

Boreal: Suggesting a fauna with preference for a cold climate.

Bunodont: A condition in which molar teeth possess rounded **cusps** and **tubercles**.

Calcareous: Made predominately of calcium carbonate, as in limestone and chalks.

Calice: The upper portion of a coral skeleton.

Calyx: A cup-shaped structure in crinoids to which the arms are attached.

Cartilage: An animal tissue that is tough but flexible.

Cast: The filling of a mold.

Catastrophist: One who believes that many features of the earth are the result of sudden widespread catastrophes rather than gradual evolutionary processes.

Caudal: Vertebrae of the tail of an animal.

Centrum: The main body of a vertebra.

Cephalon: The head of a trilobite.

Cephalothorax: The combined head and trunk parts of an arthropod.

Cervical: Vertebrae of the neck of an animal.

Chevron: A y-shaped bony arch connected to the base of the **caudal vertebrae**, designed for the attachment of tail muscles and to protect nerves and blood vessels.

Chitin: A horny substance forming all or part of the skeleton of arthropods.

Cladistics: A method of classification in which lifeforms are placed into **taxonomic** groups when they share characteristics (known as **homologies**) that are thought to indicate common ancestry.

Cold-blooded: Same as **ectotherm**.

Colony: An assemblage of connected organisms.

Comparative anatomy: Studying the structure or anatomy of living things and comparing their parts, organs and tissues to suggest groupings and relationships.

Coprolite: Fossilized animal fecal droppings.

Corallite: In corals, an individual **polyp**'s skeleton.

Correlation: In **geology** and **paleontology**, the demonstration, or attempted demonstration, of the equivalence of two or more geologic or paleontological phenomena in different geographic areas. **Fossils** are the usual method used to correlate rocks in different areas of the earth.

Cranium: The part of the skull enclosing the brain.

Creation science: Using biblical **presuppositions** as the basis for interpreting the evidences of the universe.

Cusp: A projection of the chewing surface of a tooth.

Cuticle: The hardened outer surface of the external skeleton of arthropods.

DNA: Deoxyribonucleic acid, which is constructed by combining nucleotides in long chains. Most genes consist of many thousands of nucleotides. DNA is the genetic material of living organisms and contains the information for life.

Dentary: The bone in the lower jaw that forms the side of the chin.

Denticle: A small, toothlike structure.

Dermal: Relating to the skin of an animal.

Dermal armor: The bony plates

situated in the skin of some **vertebrates**.

Detritus: Fine particles of organic matter.

Dimorphism: The existence of two different types of individuals within a **species**. One example is sexual dimorphism in animals, in which the two sexes differ in size, color, markings etc.

Dinosaur: All of the kinds of land animals that were either **ornithischian** or **saurischian**.

Disarticulated: Fossil bones found separated from the normal orientation of the living animal.

Dorsal: To the back of an animal.

Ectotherm: An animal that receives most of its body heat from its environment.

Endotherm: An animal, like a bird or mammal, that obtains most of its body heat from its own metabolism.

Erosion: Wearing away of rocks and other materials by the forces of weather such as rain, sun, wind, snow and ice.

Evidence: Something that makes another thing clear or able to be interpreted.

Evolute: Loosely coiled.

Evolution: Various models of origins that assume that all organisms on earth descended from a common ancestor by natural processes.

Exoskeleton: The hard outer casing of arthropods.

External mold: An impression of the outside of a fossil organism.

Facial suture: A line on the head of a trilobite along which splitting occurred during molting.

Fact: A reality; truth.

Family: Rank in standard **taxonomy** that lies below **order** and above **genus**.

Femur: The thigh bone of **vertebrate** animals.

Fibula: The smaller and outer of the two bones (other **tibia**) between the knee and the ankle in **vertebrate** animals.

Foot: In **invertebrates**, the base from which the organism grows, or by which it is cemented to the substrate.

Foramen: A gap, hole, opening or window in a surface or object.

Fossil: Remains or traces of life usually found embedded in **sedimentary rock**. From the Latin word *fossilis* which means "dug up."

Fossil graveyard: Bone bed.

Frond: The leaf of a fern.

Gastrolith: A stone in the stomach of an animal deliberately swallowed to aid in digestion.

Genal angle: The angle between the back and lateral margin of a trilobite's head.

Genal spine: A trilobite's cheek spine.

Genus (plural **genera**): Rank in standard taxonomy that lies below the **family** group and above the **species**.

Geologic column: A mental abstraction originally devised to attempt to explain the earth's **geology** but now used to attempt to demonstrate organic evolution in the rock record.

Geology: The study of the earth including rocks, minerals, and its surface and internal processes. Attempts to forensically determine the history of the earth is called "historical geology."

Gizzard: The part of the stomach of some animals in which food is broken up by the action of muscles and possibly **gastroliths**.

Glabella: In trilobites, the central region of the head.

Growth lines: The dividing lines between periods of growth.

Guard: A massive, bullet-shaped calcite structure in belemnites.

Hinge: The linear area along which a mollusk shell articulates.

Hinge plate: In mollusks, the portion of the valve that supports the **hinge teeth**; in brachiopods, the socket-bearing portion of the **dorsal** valve.

Hinge teeth: Articulating structures in mollusk shells.

Homology: Similarities in the structures of living things, for example the flippers of seals and the hands of people.

Humerus: The long bone of the upper arm.

Hypothesis: An unproved theory, tentatively accepted to explain certain facts.

Ichnology: The study of **trace fossils**, especially of **fossil** footprints.

Ichnogenus: A genus known only from **trace fossils**.

Ichnotaxonomy: Taxonomy developed for and applied to **trace fossils**.

Igneous rock: Rock formed by the cooling of molten material. Some rocks crystalize underground (e.g.

granite), and others at the surface (e.g. basalt).

Ilium: The largest of the three bones that make up each half of the **pelvic girdle** and are connected to the **sacral vertebrae**. The ilium, the **ischium**, and the **pubis** form the socket for the **femur**.

Index fossil: The **fossil** of an organism believed to have lived during a narrow, well-defined interval of geologic time, and used for **correlation** of rock bodies. In practice, most index **fossils** are marine **invertebrates**.

Internal mold: An impression of the inside of an organism.

Interpretation: An explanation using certain **presuppositions**.

Invertebrates: Animals without backbones.

Ischium: Bone of the **pelvis** that is directed backwards.

KT boundary: In secular **geology**, the boundary between the Cretaceous period of the Mesozoic era, and the Tertiary period of the Cenozoic era. By secular definition it also demarcates the time of the extinction of the **dinosaurs**.

Kind: A biblical category of life. God created according to kinds.

Lateral: Toward the side of an animal.

Leaflet: Segment of a compound leaf or **frond**.

Living fossil: A living plant or animal that looks identical or very similar to a fossil version of it.

Macroconch: The larger form of a shell in a species in which males and females differ in size.

Mammal: Any of a class (Mammalia) of **vertebrate** animals that nourish their young with milk secreted by mammary glands.

Mandible: Lower jaw, comprised of **dentary** bones.

Mantle: The external body wall lining the shell of some invertebrates.

Manus: Hand or front foot.

Marl: A **calcareous** mudstone.

Mass extinction: The sudden death of a large number of animal groups. The Genesis Flood was the greatest of all mass extinction events.

Matrix: The rock or sediment in which a **fossil** is embedded.

Maxilla: One of a pair of bones that form much of the skull **anterior** to the braincase and hold all of the upper teeth except the incisors.

Medial: Direction toward the middle or inside of an animal.

Megatracksite: A geographical location with large numbers of **fossil** tracks.

Metacarpals: Bones of the hand or front foot.

Metamorphic rocks: Rocks that have changed their crystalline and mineral makeup via pressure and heat without melting.

Metatarsals: Bones of the hind foot.

Microconch: The smaller form of a shell in a species in which males and females differ greatly in size.

Mold: An impression obtained from an original form.

Morphology: In **biology**, the study of form and structure.

Mummification: When preserva-tion of dead organisms is aided by severe drying out.

Mutation: A change in the genetic makeup of an organism due to exposure to things such as chemicals or radiation. Almost all mutations are harmful to any organism and evolutionists are yet to demonstrate any reality to their belief that mutations drive organic evolution. See also **natural selection**.

Naturalism: The philosophy or religion that nature (matter/energy) is the only reality, and that everything in the universe can be explained in those terms without resource to the supernatural. In the secular world "science" and "naturalism" are synonyms.

Natural selection: The mechanism proposed by Darwin to explain his concept of evolution. It describes the adaptation of organisms to their environment and explains the resulting degrees of survival and reproductive success. Creationists would allow for natural selection (first proposed by a creationist) to affect variability within a kind limited by the information originally placed in the **DNA**. Natural selection would work to reject mutations in order to save the organism rather than use the harmful mutations for progressive evolutionary improvement of some sort. The description of the process as "natural selection" is flawed in the sense that nature (or environment or ecology) is not intelligent and therefore incapable of "selecting" anything. The creationist would emphasize that adaptation of the organism to its environment is due to innate abilities of the organism, while the evolutionist must attribute intelligence to inanimate things like nature, environment, climate etc.

Neural arch: The upper portion of a **vertebra**.

Neural canal: The channel through which the spinal cord passes.

Neural spine: The blade or prong-like structure located on the **dorsal** aspect of a **vertebra**.

Nodes: Bumps or protuberances in plants, the attachment point of a leaf stem.

Nymph: An immature insect, or, in bivalves, the narrow ledge on the **hinge** behind the **umbo**.

Opisthotonic pose: The pose a dead animal assumes with its head thrown backwards, it body arched, and its tail arched upwards. Numerous **dinosaur** skeletons have been found in this pose.

Orbit: An eye socket.

Order: Rank in standard **taxonomy** that lies below class and above the **family**.

Ornithischia: A grouping of **dinosaurs** with a pelvic structure similar to that of birds. The word means "bird-hipped."

Ossicles: In **invertebrates**, the **calcareous** bodies that make up the skeleton.

Ossified: Turned to bone.

Paleontology: The study of plant and animal life from the past, including **fossils** found in the rock record.

Paradigm: A way of looking at a particular phenomenon; more encompassing than a theory or model, although sometimes used synonymously with those terms.

Peduncle: In some **invertebrates**, the stalk that supports most of the body.

Pelagic: Living in the open sea.

Pelvic girdle (Pelvis): The structure in **vertebrates** to which the **posterior** limbs are attached. It is made up of two halves, each produced by the fusion of the **ilium**, **ischium**, and **pubis**.

Permineralization: The addition of minerals to a bone during fossilization. Not all fossil bones are permineralized.

Pes: Hind foot of an animal.

Petrification: Turning to stone or rock.

Phragmocone: In belemnites and other mollusks, the conelike internal shell that is divided into chambers by **septa** and perforated by a **siphuncle**.

Plankton: Weak-swimming or passively floating animals and plants.

Pleural: Having to do with the lungs or breathing anatomy of animals.

Pleural lobes: The lateral parts of the segments of the **thorax** in trilobites.

Pleural spine: The body spine of a trilobite.

Polyp: An individual member of a coral colony.

Polystrate fossil: Any fossil that crosses two or more sedimentary layers.

Posterior: Toward the tail end of an animal or bone.

Presupposition: Something supposed or assumed beforehand.

The accumulated presuppositions of a person are the foundation for his/her **worldview**.

Process: An extension or appendage to an organism.

Proximal: At or toward the near, inner, or attached end.

Pubis: Bone of the **pelvis** that is directed forward.

Pygidium: The tail of a trilobite.

Pyrite: Gold-colored mineral composed of iron sulfide.

Quadrupedal: Using four limbs for walking.

Radiometric dating or radioisotope dating: Measuring amounts of atomic isotopes in rocks or other objects, and then assuming that this data gives the sample age.

Radius: One of the two bones that form the lower arm (other is the **ulna**).

Radula: A horny or toothlike structure located in the mouth of all mollusks, except bivalves.

Reptile: An **ectothermic**, usually egg-laying **vertebrate**, with an external covering of scales or horny plates. Snakes, lizards, and turtles are reptiles. **Dinosaurs** used to be considered reptiles but that way of thinking is not currently universally held by paleontologists.

Ribs: In **invertebrates**, raised ornamental or structural bands; in **vertebrates**, part of the thoracic skeleton.

Rugae: Wrinkles on a shell surface.

Sacral vertebrae: **Vertebrae** that are fused together and support the **pelvic girdle**.

Sandstone: A rock made from coarse mineral grains, mainly quartz.

Saurischia: A grouping of **dinosaurs** with a **pelvic** structure similar to that of modern **reptiles**. The word means "lizard-hipped."

Sauropods: The **quadrupedal** Long-Necked **dinosaurs**.

Scapula: Portion of the shoulder blade that forms part of the socket for the **humerus**, along with the coracoid.

Science: The systemized knowledge derived from observation and study. True science is not **naturalism**.

Sclerotic ring: A ring of bony plates around the eye socket in reptiles, birds and dinosaurs.

Scutes (osteoderms): Protective hard, bony, shield-like plates or scales found in some fish, reptiles and dinosaurs.

Secularism: The anti-religious view that true knowledge can only be found through rational or empirical means. Secularists live their lives with an orientation toward the present world only, as opposed to living from the biblical perspective that life is everlasting.

Sedimentary rock: Rock layers formed when sediment settles in water, and then hardens later. (**Fossils** are usually found in sedimentary rock.)

Septum (plural **septa**): A thin dividing wall in an organism.

Sicula: The skeleton of a graptolite colony.

Siltstone: A **sedimentary rock** formed from silt deposits.

Siphon: A tubular element used for the intake of water in mollusks.

Siphuncle: A tubular extension of the **mantle** passing through all chambers of shelled cephalopods. It allows for the regulation of buoyancy of these animals.

Species: A taxonomic group into which a **genus** is divided.

Spicule: A spikelike supporting structure in **invertebrates**, especially sponges.

Spinule: A small, spinelike **process**.

Spire: A complete set of **whorls** of a spiral shell.

Sternum: The breast bone of **vertebrates**.

Stipe: A branch supporting a colony of individuals, as in graptolites.

Stomach stone: A **gastrolith**.

Stratigraphy: The study of rock strata, concerned with the characters and attributes of rocks and their **interpretation** in terms of mode of origin and geological history.

Suture: A line on gastropod shells where **whorls** connect.

Symbiosis: The mutually beneficial inter-relationship between two different organisms.

Taphonomy: The study of the conditions and processes by which organisms are fossilized and preserved.

Taxonomy: The orderly arrangement of animals and plants according to their presumed natural relationships. Classical taxonomy's hierarchical classifications are (in descending order) kingdom, phylum, class, **order**, **family**, **genus**, and **species**. For evolutionists, the relationships are assumed to indicate that all life has a common ancestor.

Tectonics: Large scale deformation of the earth's crust.

Telson: In crustaceans, the last segment of the body.

Test: A hard external covering or shell.

Tetrapod: Four-legged, usually referring to **vertebrate** animals with four limbs.

Theca: In graptolites, the organic-walled tubes housing **zooids**.

Theropods: The **bipedal saurischian dinosaurs**.

Thoracic: Pertaining to the **thorax**.

Thorax: In certain arthropods, the middle of the three main divisions of the body; in vertebrates, the chest region.

Tibia: Primary bone of the lower hind leg. See also **fibula**.

Trace fossil: Fossilized tracks, trails, burrows, droppings, or tubes resulting from the life activities of animals.

Tubercle: A raised mound or bump on an organism.

Ulna: One of the two bones of the lower arm. The upper end of the ulna forms the elbow. See also **radius**.

Umbilicus: The first-formed region of a coiled shell.

Umbo (plural **umbones**): The beaklike first-formed region of a bivalve shell.

Uniformitarianism: The view that geological and biological changes in the past always occurred at the slow rate often measured today.

The principle is regularly stated as "the present is the key to the past."

Venter: In arthropods, the undersurface of the **abdomen**; in mollusks, the external, convex part of a curved or coiled shell.

Ventral: Pertaining to a surface at the interior of an animal.

Vertebra (plural vertebrae): A bone of the backbone. The vertebrae together support the body and protect the spinal cord.

Vertebrates: Animals with backbones.

Viscera: The organs within the body of an animal.

Warm-blooded: Same as "**endotherm**."

Whorl: One complete turn of a shell.

Worldview: The way a person looks at the world or his perception of reality based on his basic **presuppositions** about what is true. Every worldview, whether expressly religious or supposedly non-religious, is a belief system that begins with **presuppositions** (assumptions) held by faith.

Zooid: An individual of a colonial animal, such as in corals, graptolites and bryozoans.

Collection Index

Dinosaur Fossils:

Mammal Fossils:

No.	Description	Figure
MF10	*Poebrotherium* teeth section	193
MF11	*Hyracodon* mandible	198
MF12	*Archaeotherium* teeth section	191
MF13	*Glyptodon* scute	199

Marine Fossils:

No.	Description	Figure
MARF2	*Diplomystus* fish	153
MARF3	*Geocoma* starfish	73
MARF5	*Douvilleiceras* ammonite	57
MARF6	*Plesiosaurus* tooth	173
MARF7	*Perisphinctes* ammonite	54
MARF8	*Calymene* trilobite	88
MARF9	*Paralejuris* trilobite	86
MARF10	*Scaphites* ammonite	56
MARF11	*Dactylioceras* ammonite	49
MARF12	*Promicroceras* ammonite	53
MARF13	*Orthoceras* nautiloid	65
MARF14	*Cenoceras* nautilus	69
MARF15	*Craspedites* ammonite	48
MARF16	*Craspedites* ammonite	48
MARF17	*Pholadomya* bivalve	30
MARF18	*Cenoceras* nautilus half	69
MARF19	Plesiosaur tooth	173
MARF20	*Globidens* tooth	170
MARF21	*Platecarpus* tooth	165
MARF22	*Mosasaurus* tooth	165
MARF23	*Nothosaurus* tooth	173
MARF24	*Mortoniceras* ammonite	55
MARF25	*Knightia* fish	153
MARF26	*Desmoceras* ammonite	44-46
MARF27	*Katherinella* bivalves	29

No.	Description	Figure
MARF61	Aptychi ammonite jaws	60
MARF62	*Gryphaea* pelecypod	36
MARF63	*Dentalium* scaphopod	64
MARF64	*Chesapecten* scallop	35
MARF65	*Priscacara* fish	154
MARF66	*Eurypterus* sea scorpion	14
MARF67	*Platystrophia* brachiopod	38
MARF68	*Constellaria* bryozoan	23
MARF69	*Onchopristis* tooth	159
MARF70	*Astylospongia* sponge	24
MARF71	*Hindia* sponge	25
MARF72	*Cerastostreon* oyster	32
MARF73	*Mosasaurus* jaw	167
MARF74	*Ellipsocephalus* trilobite	96
MARF75	Crocodile scutes	178
MARF76	*Ichthyosaurus* vertebrae	161
MARF77	*Baculites* ammonite	62
MARF79	*Anadara* bivalve	31
MARF80	*Liracassis* gastropod	43
MARF81	*Eldredgeia* trilobite	97
MARF82	*Cambropallas* trilobite	93
MARF83	*Phyllograptus* graptolite	26
MARF84	*Asaphus* trilobite	99
MARF85	*Nankinolithus* trilobite	98
MARF86	*Lopha* oyster	33
MARF87	*Monograptus* graptolite	27
MARF88	*Didymograptus* graptolite	28
MARF89	*Crotalocephalus* & *Reedops* trilobites	92

Other Fossils:

No.	Description	Figure
OF1	*Hydrophilus* beetle	84

No.	Description	Figure
OF2	*Goniobasis* gastropod	41
OF3	Birch leaf	210
OF4	*Salix* leaf Oregon	211
OF5	*Metasequoia* leaves	216
OF6	Crane flies	82
OF7	*Alethopteris* fronds	218
OF8	Petrified wood section w/rings	207
OF9	*Ginkgo* petrified wood	206
OF10	*Lepidodendron* bark, lycopod bark	219, 220
OF11	*Eryops* claw and toe	186
OF12	*Neuropteris* leaflet	221
OF13	Insects in amber	85
OF14	*Araucarioxylon* wood	208
OF15	*Annularia* leaf whorl	223
OF16	*Pecopteris* frond	222
OF17	*Calamites* stem	224
OF18	Trilobite tracks	232
OF19	*Ginkgo* leaf	205
OF20	*Salix* leaf Idaho	212
OF21	*Salix* leaf Utah	213
OF22	*Zelkova* leaf	214
OF23	*Toredo* wood	239
OF24	Worm burrow cast in matrix	241
OF25	Worm burrow cast no matrix	242
OF26	Flower	226
OF27	*Sequoia* seed cone	217
OF28	*Acer* leaf	215
OF29	Raindrops and ripples	235, 236
OF30	Parasitic wasp	83

Dinosaur Replicas:

No.	Description	Figure
DR1	*Tyrannosaurus* tooth	118
DR2	*Triceratops* horn	131
DR4	*Tyrannosaurus* claw	120
DR5	Duck-Billed dinosaur skin impression	130
DR6	Tyrant Bipedal dinosaur egg	233
DR7	Dinosaur footprint Massachusetts	229
DR8	*Coelophysis* skull	148
DR9	*Nanosaurus* foot	120
DR11	Dinosaur coprolite	227
DR12	Horn-Faced dinosaur skin impression	130
DR13	*Brachiosaurus* toe bone	146
DR14	*Megaraptor* claw	112
DR15	*Archaeopteryx* Berlin specimen	107
DR16	Duck-Billed dinosaur egg	233
DR17	*Compsognathus* specimen	109
DR19	*Tyrannosaurus* tooth with root	118
DR20	*Allosaurus* furcula	113
DR21A	*Archaeopteryx* Eichstatt specimen	108
DR22A	*Archaeopteryx* skull	103, 104
DR22V	*Velociraptor* skull	103, 104
DR23	*Magnoavipes* footprint	230
DR24	*Tyrannosaurus* furcula	113
DR25	*Othnielia* foot	120
DR26	*Camarasaurus* tooth	143
DR28	*Edmontosaurus* vertebra	126
DR29	*Edmontosaurus* tooth	128
DR30	*Edmontosaurus* tooth row	128
DR31	*Stegosaurus* tail spike	140
DR32	*Allosaurus* tooth	118
DR33	*Stegosaurus* tail plate	142
DR34	*Allosaurus* femur	121
DR35	*Allosaurus* radius bones	123

Other Replicas:

No.	Description	Figure
OR1	*Protolindenia* dragonfly	80
OR4	*Mesolimulus* horseshoe crab	13
OR5	*Coelacanth* fish	156
OR10	*Gallinuloides* bird	179
OR11	Pigeon skeleton (actual)	105
OR12	*Mesohippus* skull	194
OR13	*Hesperocyon* skull	197
OR14	*Hoplophoneus* skull	196
OR15	Peccary skull (actual)	195
OR17	*Poebrotherium* skull	193
OR18	*Trionyx* turtle	184
OR19	*Pliosaurus* tooth	171
OR20	*Sarcosuchus* tooth	174
OR21	*Pterodactylus kochi* pterosaur	180
OR22	*Ginkgo biloba* leaf	204
OR23	*Pterodactylus antiquus* pterosaur	181
OR24	*Rhamphorhynchus* pterosaur	182
OR25	*Sarcosuchus* scute	177
OR26	*Dicranurus* trilobite	95
OR27	*Ichthyosaurus* flipper	164
OR28	*Eryon* lobster	17
OR29	*Propelodytes* frog	185
OR30	*Icaronycteris* bat	188
OR31	*Lepisosteus* fish	158
OR32	*Walliserops* trilobite	100

Living Fossil Index

Living Fossil: A fossil animal or plant that looks very similar to a living organism.

The recovery and identification of hundreds of living fossils from the rock record fits perfectly into the biblical worldview. However, it is a real problem for the secular worldview that presupposes hundreds of millions of years of macroevolution. This book, *Fossils: Description & Interpretation* includes just a sampling of these living fossils which are listed below.

Living Fossils – Invertebrate Animals:

Description	Figure(s)	Description	Figure(s)
Bivalve clams	29-31	Oysters	32-33
Corals	78-79	Parasitic wasp	83
Crane fly	82	*Pecten* scallops	34-35
Crinoid	18	Sand dollars	74-75
Dragonfly	80	Scaphopod	64
Gastropods	39, 41, 43	Shrimp	16
Horseshoe crab	13	Sponges	24-25
Insects in amber	85	Starfish	72-73
Lobster	17	Water beetle	84
Mayfly	81		
Nautilus	69-71		

Living Fossils – Vertebrate Animals:

Description	Figure(s)	Description	Figure(s)
Bat	188	Peccary	195
Cat	196	Perch fish	154
Coelacanth fish	156	Quail	179
Crocodile	174, 176-178	Sawfish	159
Dog	197	Shark	202
Frog	185	Turtle	184
Garfish	158	Whale	201
Herring fish	153		

Living Fossils – Plants:

Description	Figure(s)	Description	Figure(s)
Elm leaf	214	Maple leaf	215
Fern fronds	218	*Metasequoia* leaves	216
Flower	226	*Sequoia* seed cone	217
Ginkgo leaves	204-206	Willow leaves	209-213

Subject and Name Index

Knightia 154-155

L

La Brea Tar Pits 87-88
Lagerstatte 84, 154
Lebanon 19, 84, 153
Lepidodendron 216-217
Lepisosteus 158-159
Leviathan 173, 175
Liracassis 47
Lithe, Fast Running 152
Living fossils (See Living Fossil Index)
Lobster 20-21
Lopha 37-38
Lycopods 216-218
Lyell, Charles 161
Lyme Regis 56, 72, 161-162

M

Macrofossils 5
Madagascar 37-38, 48-51, 60, 64, 73, 77
Magnoavipes 228
Mammoths 5
Mandible 166-167, 187, 190-196, 245
Maple 207, 212
Marrow, Bone 142-144
Mass mortality 53, 93
Mastodons 5
Mayfly 84
Mazon Creek 218-220
Megalodon 198-200
Megaraptor 119-120
Merycoidodon 185-187
Mesohippus 191
Mesolimulus 16-17
Messel Pit 159, 182
Metasequoia 213-214
Microfossils 5

Miohippus 191
Mioplosus 154
Monograptus 30-31
Monticules 26-27
Morocco 21, 61, 68-71, 75, 81, 90, 92, 94, 96-97, 99, 102, 105, 160, 165-171
Morrison formation 147, 226
Morrison Natural History Museum 150
Mortoniceras 58-59
Mosaic 110, 120, 122
Mosasaurus 165-168
Museum of the Rockies 136

N

Nankinolithus 102-103
Nanofossils 5
Nanosaurus 126
Nautiloids 49, 68-71
Nautilus 49, 69-70, 72-74
Neural spine 133, 247
Neuropteris 218-219
Noah 5, 21, 57, 86, 117, 142, 152, 176, 187, 190, 198, 202, 205, 209, 229, 236, 272
Noah's ark 2, 42, 130
Nothosaurus 171-172
Numbering system 8

O

Oak 207
Onchopristis 159-161
Ophiuroids 76
Opisthotonic pose 117, 246
Oreodont 12-13, 185-187
Origin of Species, The 167, 271
Orohippus 191
Orthoceras 68-71
Osteoderms 175-176, 197, 247
Othnielia 126

Oyster 36-38

P

Paralejuris 89-90
Parasitic wasp 86-87
Patagonia 119
Pathology 130
Peccaries 192-193
Pecopteris 219-220
Pecten 9, 38-40
Pentremites 23
Perchoerus 192
Perisphinctes 57
Permineralization 13, 50, 246
Petoskey stones 82
Petrified Forest National
    Park 205-206
Petrified wood 203-205
Phacops 94-95
Phillipsastrea 81
Pholadomya 34
Phyllograptus 29
Pierre Shale 65-66, 168
Pigeon 110-112, 118
Plate-Backed 145-146
Platecarpus 165
Platypus 110, 112, 122
Platystrophia 42-43
Plesiosaurus 171-172
Pliohippus 191
Pliosaurus 169-170
Poebrotherium 190
Polystrate fossils 159, 181, 247
Poplar 207
Porifera 28
Presuppositions 3-4, 6, 38, 40, 61,
    66, 71, 79, 108, 123, 163-164,
    172, 191, 196, 223, 247
Priscacara 155-156
Promicroceras 56
Propelodytes 181-182

Protohippus 191
Protolindenia 83
Pteranodon 179
Pterodactylus 177-179
Pygurus 77
Pyrite 41-42, 247

R

Radiocarbon dating 89
Radius bone 129-130, 247
Raindrops 232-234
Rainbow Wood 203, 205
Redwall limestone 71
Reedops 95-96
Replicas 4, 8, 14-15, 108, 129,
    256-257
Resin 88-89, 241
Rhamphorhynchus 179-180
Romans, book of 42, 64, 162, 223
Royal Tyrrell Museum 184, 189
Running Rhino 195

S

Sahara Desert 21, 68-69
Salix 208-210
Sand dollar 77-78
Sarcosuchus 172-176, 197
Sawfish 159-161
Scallops 39-40
Scaphites 59-60
Scaphopods 67
Scutes 174-176, 196-197, 247
Scyphocrinus 21
Sea scorpion 17-18
Secular Worldview 1, 7, 159, 259
Septae 49, 51
Sequoia 213-214
Shipworms 237
Silica 46, 205-206
Sin 2, 7, 130, 137, 189
Siphuncle 49, 70, 72, 74, 246, 248

Sicula 32, 247
Skin imprints 136-137
Skulls 12-13, 15, 109-110, 112-113,
    115, 122-123, 137, 141, 147-148,
    151-152, 173, 177-181, 183, 186-
    196, 242, 245
Smithsonian Museum 51, 66, 174
Solnhofen 11, 16, 19-20, 63, 76,
    83-84, 113-116, 122, 155, 157,
    177-179, 182
South Dakota Museum of
    Geology 132, 168
Spicules 27, 248
Spirifer 41-42
Sponges 27-29
Starfish 74-76
Stegosaurus 145-147
Stenaster 74-75
Sternum 111, 248
Stewartville formation 238
Stipes 29-32, 248
Styracosaurus 141
Suture lines 49-52, 66, 248

T
Tayassu 192
Teeth 13, 31, 37-38, 77, 109, 124-
    126, 135, 138, 147-149, 151, 153,
    160, 165-172, 176, 181, 186-198,
    200, 209-210, 241, 244
Tegana formation 160
Theca 23, 29-32, 248
Thermopolis specimen 114, 117-118
Tonna 47
Toredo wood 236-237
Torosaurus 141
Toyen formation 29
Transitional fossils 3, 109-110, 122,
    272
Triceratops 137-141
Trilobite eyes 89-95, 99-104
Trilobite legs 100

Trilobites 89-106
Trinucleidae 103
Trionyx 108-181
Turritella 43-46
Turtles 180-181, 197
Tusk shell 67
Tylosaurus 165-166
Tympanic bulla 198
Tyrannosaurus 120, 122-126
Tyrant Bipedal 121-127, 228, 231

U
Uniformitarianism 4, 79, 234, 238,
    249
Upper Silesia formation 216-217
Utahraptor 120

V
Variation within kinds 3-4, 7, 13,
    21, 27, 36-37, 40-41, 43, 77, 81,
    89, 96, 99, 103-104, 120, 123,
    125, 128, 138, 155, 165, 178,
    184-185, 191, 193-195, 198, 202-
    203, 211, 217
Velociraptor 109-110, 120
Vertebrae 112, 133, 140, 161-162,
    242, 245, 247, 249
Vilpovitsy Quarry 104

W
Walliserops 105-106
Walnut formation 36
Water ripples 234-235
Wellnhofer, Peter 110, 113, 119,
    122
Whales 164, 198-199
Wheeler Shale 93-94, 207
White River Badlands 13, 188-192
White River formation 14, 186-187,
    193, 195-196
Willow 207-210
Worm burrows 238-239

Wyoming Dinosaur Center 22, 44,
  53, 180

Author's Biography

J.D. Mitchell is a retired registered professional engineer and has a Bachelor of Science in Mechanical Engineering from the University of Washington. He has also completed his Master of Biblical Studies in Biblical Creation Apologetics from Master's Graduate School of Divinity.

He is Executive Director of the Institute for Creation Science that meets monthly in Portland, Oregon. He does creation science research and writes and speaks regarding the creation versus evolution controversy as a part of his creation ministry Creation Engineering Concepts. He is also a member of the Design Science Association of Oregon and the Creation Research Society.

J.D. has been studying the scientific and biblical evidence regarding the origins controversy since 1984 when he was converted from a theistic evolutionist to a born again Christian. *Fossils: Description and Interpretation* is the sixth book he has published on origins topics.

Dr. Jerry Bergman

Fossils: Description & Interpretation – Within a Biblical Worldview, by author J.D. Mitchell, is a tour de force of nearly 250 of the common fossils found in many parts of the world. It includes not only a description of each fossil, but color pictures from the author's personal collection to help reveal their beauty and details. In his 1859 book, *The Origin of Species,* Darwin lamented that, if it were not for the "imperfection of the fossil record," (the title of chapter 10), his theory would be proven. Darwin claimed that the fossil record will make or break his theory, and Mitchell's book appears to break his theory. Darwin wrote "Geology assuredly does not reveal any such finely graduated organic chain; and this, perhaps, is the most obvious and gravest objection which can be urged against my theory. The explanation lies, as I believe, in the extreme imperfection of the geological record."

It was expected by Darwin and others that these gaps would be filled in with more research, yet they were not filled in, either in his day nor in ours. The fact is, "Darwin could not point to a single example where fossils in successive geological strata showed evolution from one species to another."* Darwin felt confident that, as new fossils were uncovered, evidence for his theory would eventually be revealed. One reason, Darwinists claim, that the gaps have not been filled in is because fossils are preserved only under very unusual circumstances. Nonetheless, the museums and private collections contain millions of fossils, and these fossils still do not show the expected record of transitional forms.

Scientists now recognize that there could be as many as ten million different species, and paleontologists have confirmed that about a million kinds of animal and plant species exist. Thus, if as scientists estimate, from 50 to 100 transitions would be required to evolve a single new species, trillions of transitional fossils must have existed. If even only one out of a million of these transitions were preserved in the fossil record, we would have good evidence of the gradual evolution of all life, but the hoped for fossil "links" have not been found to exist.

The major way fossils are preserved is rapid burial, such as by a catastrophic flood. The vast majority of organisms that die are rapidly consumed by other life forms, from bacteria to maggots to vultures. Mitchell shows that the fossils in his collection, and many of those outside of it that he has photographed, were preserved by rapid burial as you would expect in a cataclysmic flood. He then argues that the flood of Noah fits the bill perfectly. The explanation for the existence of the millions of fossils so far uncovered, some exceptionally well preserved, is they are evidence for a catastrophic flood such as described in Genesis.

---

* Pearson, Paul, *Nature Debates*, "The Glorious Fossil Record," Nov. 19, 1998, p.1.

www.ingramcontent.com/pod-product-compliance
Lightning Source LLC
Chambersburg PA
CBHW042309210326
41598CB00041B/7327